Plows, Plagues, and Petroleum

Plows, Plagues, and Petroleum

HOW HUMANS TOOK CONTROL OF CLIMATE

With a new afterword by the author

William F. Ruddiman

PRINCETON UNIVERSITY PRESS

PRINCETON AND OXFORD

Copyright © 2005 by Princeton University Press

Published by Princeton University Press, 41 William Street, Princeton,
New Jersey 08540

In the United Kingdom: Princeton University Press, 6 Oxford Street, Woodstock,
Oxfordshire OX20 1TW

First printing, 2005
First paperback printing, 2007
First Princeton Science Library edition, with a new afterword, 2010
ISBN: 978-0-691-14634-8

The Library of Congress has cataloged the cloth edition of this book as follows

Ruddiman, W. F. (William F.), 1943–
    Plows, Plagues, and Petroleum : how humans took control of climate /
    William F. Ruddiman.
        p.   cm.
    Includes bibliographical references and index.
    ISBN 0-691-12164-8 (cloth : alk. paper)
        1. Climatology.   2. Climatic changes.   3. Global temperature changes.
    4. Greenhouse effect, Atmospheric.   I. Title.

QC981.R763 2005
363.738′74—dc22   2004062444

British Library Cataloging-in-Publication Data is available

This book has been composed in Garamond

Printed on acid-free paper. ∞

press.princeton.edu

Printed in the United States of America

3   5   7   9   10   8   6   4   2

For Ginger, Alysha, and Dustin

CONTENTS

LIST OF ILLUSTRATIONS

FIGURES

TABLES

PREFACE

THE RESEARCH THAT LED to this book began when I was a professor in the Department of Environmental Sciences at the University of Virginia, following my earlier career at Lamont-Doherty Observatory of Columbia University. I was able to interest undergraduate student Jonathan Thomson in a term-length research project that explored a mystery that had puzzled me earlier: the fact that methane concentrations in the atmosphere have risen for the last 5,000 years when everything I knew about the natural controls of methane predicted that a drop should have occurred instead. In 2001 we published a joint paper in the peer-reviewed literature attributing the anomalous methane trend to human activities.

After retiring early in 2001, I began working at home on a similar mystery—a rise in carbon dioxide concentrations that occurred during the last 8,000 years even though natural factors (as I understood them) again predicted a drop. The 2003 publication of a paper summarizing my new hypothesis of early human effects on climate coincided with a lecture at the annual American Geophysical Union meeting in San Francisco. More recently I have talked to university audiences far larger than earlier in my career. People are interested in and stimulated by this hypothesis, but acceptance of radically new ideas does not come quickly in science; the community is still sorting out their reactions, and the ultimate judgment concerning this hypothesis lies in the future.

Because the detective work that forms the central theme of this book draws on my four-decade experience in climate science, thanking everyone involved in this hypothesis (and thus this book) would be an endless task. Still, some thanks stand out from the rest. My background in understanding the impact of orbital variations on Earth's climate came from working with the CLIMAP group, among which John Imbrie was the guiding light and Nick Shackleton and Jim Hays were key members. My understanding of the effects of ice sheets and tropical monsoons on climate came from working with the COHMAP group, with John Kutzbach as primary "teacher" and Tom Webb, Herb Wright, and Alayne Street as important members. Work with John Kutzbach on the climatic effects of uplift of the Tibetan Plateau taught me about the uses and limitations of climate models. Collaborations with colleague Andy McIntyre and with my former graduate students Alan Mix, Ned Pokras, Glenn Jones, Maureen Raymo, and Peter deMenocal also expanded my horizons. In recent years, discussions with and/or publications by many others broadened my understanding of several aspects of climate and human history. Andre Berger, Wally Broecker, Jared Diamond, Richard Houghton, Mike Mann, Neil Roberts, Pieter Tans, Jim White, and Michael Williams deserve special mention.

I thank Bob Smith for the illustrations, W. H. Freeman for approval to adapt several figures from my textbook *Earth's Climate*, and Jack Repcheck and Holly Hodder for steering me through some of the tangles of the publishing world. Ingrid Gnerlich did a fine job moving this manuscript through the review system at Princeton University Press. For other contributions to this book, I thank Anita O'Brien for careful copy editing, and Dale Cotton for overseeing the production process. My retirement funds in TIAA/CREF provided financial support of this effort.

PART ONE

What Has Controlled Earth's Climate?

WHAT HAS CONTROLLED EARTH'S CLIMATE?

IMAGINE EARTH VIEWED from a satellite. Blue oceans cover more than two-thirds of the planet and brown or green land the rest. White ice sheets over a mile thick bury a small fraction of the land (Antarctica and Greenland). Whitish sea ice forms a cap a few feet thick over the polar oceans, and its seasonal fluctuations in the two hemispheres occur at exactly opposite tempos (one large when the other is small). Surrounding everything is a thin blue envelope of atmosphere with swirls of clouds.

In comparison to these fundamental and massive parts of the natural climate system, the largest structures built by humans are insignificant to or even undetectable by the unaided eye. Pyramids, dams, and roads are invisible from space without high-powered telescopes. On the side of Earth lying in the dark of night, even the brightly lit cities are just tiny islands of light.

From this perspective, the possibility that humans could have any major impact on the workings of these vast parts of the climate system sounds ridiculous. How could we possibly cause changes in the size of these immense regions of blue and green and white? Yet we are. No credible climate scientist now doubts that humans have had an effect on Earth's climate during the last two centuries, primarily by causing increases in the concentrations of greenhouse gases like carbon dioxide and methane in the atmosphere. These gases trap radiation emitted from Earth's surface after it has been heated by the Sun, and the added heat retained in Earth's atmospheric envelope makes its climate warmer.

Because increases in both greenhouse gases and Earth's temperature during the last century have been measured, the so-called global warming debate is not about whether humans are warming climate or whether we will warm climate in future decades—we are warming it, and we will warm it more in future decades as greenhouse-gas concentrations rise.

The only issue under serious debate is: By how much? Will we make Earth's climate only slightly warmer, a change that might be hardly noticeable? Or will we alter the climate system in much more extensive ways, for example by melting most of that white sea-ice cover near the North Pole and turning the Arctic to an ocean blue? For now, the answer to this question of "how-much" is not so clear.

Another part of the global-warming debate is whether these changes will be "good" or "bad." This question has many answers, all of which turn on the value system of the person asking it. The world is complicated; no single answer of good or bad is sufficient when the many complexities of such an issue are taken into account. But most of the story this book has to tell is not about highly charged political or media debates under way today and forgotten a few years hence. The focus here is on what we can learn from the past.

For most of the time that human beings and our recognizable ancestors lived on Earth, we did not affect climate. Few in number, and moving constantly in search of food and water, our Stone Age predecessors left no permanent "footprints" on the landscape for several million years. Throughout this immensely long span of time, climate changed for natural reasons, primarily related to small cyclical changes in Earth's orbit around the Sun. Nature was in control of climate.

But the discovery of agriculture nearly 12,000 years ago changed everything. For the first time, humans could live settled lives near their crops, rather than roaming from area to area. And gradually, the improved nutrition available from more dependable crops and livestock began to produce much more rapid increases in population than had been possible in the earlier hunting-and-gathering mode of existence. As a result, the growing human settlements began to leave a permanent footprint of ever-increasing size on the land.

If you could watch a time-lapse film showing Earth's surface since agriculture began, you would see a subtle but important change spread across southern Eurasia during the last several thousand years. In China, India, southern Europe, and northernmost Africa, you would see darker shades of green slowly turning a lighter green or a greenish brown. In these areas, the first villages, towns, and cities were being built, and vast areas of dark-green forest were slowly being cut for agriculture, cooking, and heating, leaving behind the lighter-green hues of pastures or the green-brown of croplands.

Until very recently, scientists thought that humans first began altering climate some 100 to 200 years ago, as a direct result of changes brought about by the gassy effusions of the Industrial Revolution. But here I propose a very different view: the start of the switch-over from control of climate by nature to control by humans occurred several thousand years ago, and it happened as a result of seemingly "pastoral" innovations linked to farming. Before we built cities, before we invented writing, and before we founded the major religions, we were already altering climate. We were farming.

Chapter One

CLIMATE AND HUMAN HISTORY

MOST SCIENTISTS ACCEPT the view that human effects on global climate began during the 1800s and have grown steadily since that time. The evidence supporting this view looks quite solid: two greenhouse gases (carbon dioxide, or CO_2, and methane, or CH_4) that are produced both in nature and by humans began unusual rises like the pattern shown in figure 1.1A. Both the rate of change and the high levels attained in the last 100 to 200 years exceed anything observed in the earlier record of changes from ancient air bubbles preserved in ice cores. Because greenhouse gases cause Earth's climate to warm, these abrupt increases must have produced a warming.

But one aspect of the evidence shown in figure 1.1A is deceptive. Magicians use a form of misdirection in which flashy movements with one hand are used to divert attention from the other hand, the one slowly performing the magic trick. In a sense, the dramatic change since 1850 is exactly this kind of misdirection. It distracts attention from an important rise in gas concentrations that was occurring during the centuries before the 1800s. This more subtle change, happening at a much slower rate but extending very far back in time, turns out to be comparably important in the story of humanity's effects on climate.

I propose that the real story is more like the one shown in figure 1.1B. Slower but steadily accumulating changes had been underway for thousands of years, and the total effect of these earlier changes nearly matched the explosive industrial-era increases of the last century or two. Think of this as like the fable of the tortoise and hare: the hare ran very fast but started so late that it had trouble catching the tortoise. The tortoise moved at a slow crawl but had started early enough to cover a lot of ground.

The tortoise in this analogy is agriculture. Carbon dioxide concentrations began their slow rise 8,000 years ago when humans began to cut and burn forests in China, India, and Europe to make clearings for croplands and pastures. Methane concentrations began a similar rise 5,000 years ago when humans began to irrigate for rice farming and tend livestock in unprecedented numbers. Both of these changes started at negligible levels, but their impact grew steadily, and they had a significant and growing impact on Earth's climate throughout the long interval within which civilizations arose and spread across the globe.

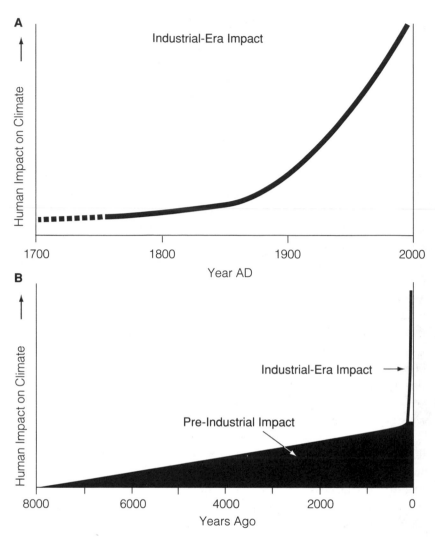

1.1. Two views of the history of human impacts on Earth's climate and environment. A: Major impacts began during the industrial era (the last 200 years). B: The changes of the industrial era were preceded by a much longer interval of slower, but comparably important, impacts.

For most people (including many scientists), the natural first reaction to this claim of a very early human impact on climate is disbelief. How could so few people with such primitive technologies have altered Earth's climate so long ago? How do we know that the "tortoise" version shown in figure 1.1B is correct? Convergent evidence from two areas of scientific research in which major revolutions of knowledge have occurred in the last half century—climatic history and early human history—provides the answer to these questions and the demonstration of an early human impact on climate.

When I started my graduate student career in the field of climate science almost 40 years ago, it really was not a "field" as such. Scattered around the universities and laboratories of the world were people studying pollen grains, shells of marine plankton, records of ocean temperature and salinity, the flow of ice sheets, and many other parts of the climate system, both in their modern form and in their past manifestations as suggested by evidence from the geologic record. A half-century before, only a few dozen people were doing this kind of work, mostly university-based or self-taught "gentleman" geologists and geographers in Western Europe and the eastern United States. Now and then, someone would organize a conference to bring together 100 or so colleagues and compare new findings across different disciplines.

Today, this field has changed beyond recognition. Thousands of researchers across the world explore many aspects of the climate system, using aircraft, ships, satellites, innovative chemical and biological techniques, and high-powered computers. Geologists measure a huge range of processes on land and in the ocean. Geochemists trace the movement of materials and measure rates of change in the climate system. Meteorologists use numerical models to simulate the circulation of the atmosphere and its interaction with the ocean. Glaciologists analyze how ice sheets flow. Ecologists and biological oceanographers investigate the roles of vegetation on land and plankton in the ocean. Climatologists track trends in climate over recent decades. Hundreds of groups with shorthand acronyms for their longer names hold meetings every year on one or another aspect of climate. I am certain there are now more groups with acronyms in the field of climate science than there were people when I began.

Studies of Earth's climatic history utilize any material that contains a record of past climate: deep-ocean cores collected from sea-going research vessels, ice cores drilled by fossil-fuel machine power in the Antarctic or Greenland ice sheets or by hand or solar power in mountain glaciers; soft-sediment cores hand-driven into lake muds; hand-augered drills that extract thin wood cores from trees; coral samples drilled from tropical reefs. The intervals investigated vary from the geological past many tens of millions of years ago to the recent historical past and changes occurring today.

These wide-ranging investigations have, over the last half-century or so, produced enormous progress in understanding climate change on every scale. For intervals lying in the much more distant past, tens or hundreds of millions of years ago, changes in global temperature, regional precipitation, and the size of Earth's ice sheets have been linked to plate-tectonic reorganizations of Earth's surface such as movements of continents, uplift and erosion of mountains and plateaus, and opening and closing of isthmus connections between continents. Over somewhat shorter intervals, cyclic changes in temperature, precipitation, and ice sheets over tens of thousands of years have been linked to subtle changes in Earth's orbit around the Sun, such as the tilt of its axis and the shape of the orbit. At still finer resolution, changes in climate over centuries or decades have been tied to large volcanic explosions and small changes in the strength of the Sun.

Some scientists regard the results of this ongoing study of climate history as the most recent of four great revolutions in earth science, although advances in understanding climate have come about gradually, as in most of the earlier revolutions. In the 1700s James Hutton concluded that Earth is an ancient planet with a long history of gradually accumulated changes produced mainly by processes working at very slow rates. Only after a century or more did Hutton's concept of an ancient planet displace the careful calculations of an archbishop in England who had added up the life spans of the patriarchs in the Bible and calculated that Earth was formed on October 26 in 4004 B.C. Today chemistry, physics, biology, and astronomy have all provided critical evidence in support of the geology-based conclusion that our Earth is very old indeed, in fact several *billions* of years old.

In 1859 Charles Darwin published his theory of natural selection, based in part on earlier work showing that organisms have appeared and disappeared in an ever-changing but well-identified sequence throughout the immense interval of time for which we have the best fossil record (about 600 million years). Darwin proposed that new species evolve as a result of slow natural selection for attributes that promote reproduction and survival. Although widely accepted in its basic outline, Darwin's theory is still being challenged and enlarged by new insights. For example, only recently has it become clear that very rare collisions of giant meteorites with Earth's surface also play a role in evolution by causing massive extinctions of most living organisms every few hundred million years or so. Each of these catastrophes opens up a wide range of environmental niches into which the surviving species can evolve with little or no competition from other organisms (for a while).

The third great revolution, the one that eventually led to the theory of plate tectonics, began in 1912 when Alfred Wegener proposed the concept of continental drift. Although this idea attracted attention, it was widely rejected in North America and parts of Europe for over 50 years. Finally, in the late 1960s, several groups of scientists realized that marine geophysical data that had been collected for decades showed that a dozen or so chunks of Earth's crust and outer mantle, called "plates," must have been slowly moving across Earth's surface for at least the last 100 million years. Within 3 or 4 years, the power of the plate tectonic theory to explain this wide range of data had convinced all but the usual handful of reflex contrarians that the theory was basically correct. This revolution in understanding is not finished; the mechanisms that drive the motions of the plates remain unclear.

As with the three earlier revolutions, the one in climate science has come on slowly and in fact is still under way. Its oldest roots lie in field studies dating from the late 1700s and explanatory hypotheses dating from the late 1800s and early 1900s. Major advances in this field began in the late 1900s, continue today, and seem destined to go on for decades.

Research into the history of humans is not nearly as large a field as climate science, but it attracts a nearly comparable amount of public interest. This field, too, has expanded far beyond its intellectual boundaries of a half-century ago. At that time, the fossil record of our distant precursors was still extremely meager. Humans and our precursors have always lived near sources of water, and watery soils contain acids that dissolve most of the bones overlooked by scavenging animals. The chance of preservation of useful remains of our few ancestors living millions of years ago is tiny. When those opposed to the initial Darwinian hypothesis of an evolutionary descent from apes to humans cited "missing links" as a counterargument, their criticisms were at times difficult to refute. The gaps in the known record were indeed immense. Now the missing links in the record of human evolution are at most missing minilinks. Gaps that were as much as a million years in length are generally now less than one-tenth that long, filled in by a relatively small number of anthropologists and their assistants doggedly exploring outcrops in Africa and occasionally stumbling upon fossil skeletal remains.

Suppose that skeletal remains are found in ancient lake sediments sandwiched between two layers of lava that have long since turned into solid rock (basalt). The basalt layers can be dated by the radioactive decay of key types of minerals enclosed within. If the dating shows that the two layers were deposited at 2.5 and 2.3 million years ago, respectively, then the creatures whose remains were found in the lake sediments sandwiched in between must have lived within that time range. With dozens of such dated skeletal remains found over the last half-century, the story of how our remote precursors changed through time has slowly come into focus.

Even though the details of the pathway from apes to modern humans still need to be worked out, the basic trend is clear, and no credible scientist that I know of has any major doubts about the general sequence. Creatures intermediate between humans and apes (australopithecines, or "southern apes") lived from 4.5 to 2.5 million years ago, around which time they gave way to beings (the genus *Homo*, for "man") that we would consider marginally human, but not fully so. Today anthropologists refer to everything that has followed since 2.5 million years ago as the hominid (or hominine) line. By 100,000 years ago, or slightly earlier, fully modern humans existed. This long passage was marked by major growth in brain size; progressively greater use of stone tools for cutting, crushing, and digging; and later by control of fire.

Knowledge of the more recent history of humans has increased even more remarkably. Decades ago the field of archeology was focused mainly on large cities and buildings and on the cultural artifacts found in the tombs of the very wealthy; today this field encompasses or interacts with disciplines such as historical ecology and environmental geology that explore past human activities across the much larger fraction of Earth's surface situated well away from urban areas. Radiocarbon dating (also based on radioactive decay) has made it possible to place even tiny organic fragments with a time framework. The development of cultivated cereals in the Near East nearly 12,000 years ago and their spread into previously forested regions of Europe from 8,000 to 5,500 years ago can be dated from trace amounts of crops found in lake sediments. On other research fronts, archeologists unearthing mud-brick and stone foundations of houses have been able to estimate population densities thousands of years ago. Others examining photos taken from the air in early morning at low sun angles find distinct patterns of field cultivation created by farmers centuries before the present. Geochemists can tell from the kind of carbon preserved in the teeth and bones of humans and other animals the mixture of plants and animals they ate. From these and other explorations, the developing pattern of human history over the last 12,000 years has come into much sharper focus.

Because both of these research fields—climatic and human history—concentrate on the past, they have much in common with the field of crime solving. Imagine that a breaking and entering and a murder have been committed. The detectives arrive and examine the crime scene, searching for evidence that will point to the guilty person. How and when did the criminal enter the house? Was anything stolen? Were muddy footprints or fibers or other evidence left behind? Based on all the evidence, and the modus operandi of the possible perpetrators, the detectives gradually zero in on the identity of the criminal. Was the crime the work of a family member, an outsider who knew the family, or a complete stranger? A list of possible suspects emerges, the detectives check out where they were at the time of the murder, and a primary suspect is identified.

By analogy, students of climate and human history also arrive on the scene after the event has occurred, but in this case hundreds, thousands, or even many millions of years later. And, as in the crime scene, the first thing these scientists encounter is evidence that something of importance has happened. Twenty thousand years ago, an ice sheet more than a mile high covered the area of the present-day city of Toronto. Ten thousand years ago, grasslands with streams and abundant wildlife existed in regions now covered by blowing sand in the southern Sahara Desert.

Natural curiosity drives scientists to wonder how such striking changes could have happened, and for some scientists this process of wondering leads to hypotheses that are first attempts at explanations. Soon after a major discovery is made, other scientists challenge the initial hypothesis or propose competing explanations. Over many years and even decades, these ideas are evaluated and tested by a large community of scientists. Some of the hypotheses are found to be inadequate or simply wrong, most often because additional evidence turns out to be inconsistent with specific predictions made in the initial hypotheses. If any hypothesis survives years of challenges and can explain a large amount of old and new evidence, it may become recognized as a theory. Some theories become so familiar that they are invoked almost without conscious thought and called paradigms. But even the great paradigms are not immune from continual testing. Science takes nothing for granted and draws no protective shield around even its time-honored "successes."

Only rarely do scientists studying climate history manage to isolate one causal explanation for any specific piece of evidence. By analogy to a crime scene, the detectives might be lucky enough to find totally diagnostic and incriminating evidence near the murder victim or at the point of the break-in, such as high-quality fingerprints or blood samples with DNA that match evidence from a suspect. If so, the perpetrator of the crime is convicted (unless the prosecutors are totally incompetent). In climate science, the explanation for an observation (the presence of ice sheets where none exist today, or of ancient streambeds in modern-day deserts) more commonly ends up with several contributing factors in plausible contention.

But sometimes nature can be more cooperative in revealing cause-and-effect connections. The changes in Earth's orbit mentioned earlier occur at regular cycles of tens of thousands of years. These same cycles have occurred during many of Earth's major climatic responses, including changes in the size of its ice sheets and in the intensity of its tropical monsoons. Because "cycles" are by definition regular in both length (duration in time) and size (amplitude), they are inherently predictable. This gives climate scientists like me a major opportunity. We can look at past records of climate and see where and when the natural cycles were behaving "normally," but if we then find a trend developing that doesn't fit into the long-term "rules" set by the natural system, we are justified in concluding that the explanation for this departure from the norm cannot be natural.

Several years ago, just before I retired from university life, I noticed something that didn't seem to fit into what I knew about the climate system. What bothered me was this: the amount of methane in the atmosphere began going up around 5,000 years ago, even though everything I had learned about the natural cycles told me it should have kept going down. It has occurred to me since then that this was like an early scene in every episode of Peter Falk's *Columbo* television series when he has just begun to investigate a recently committed crime. After he finishes an initial talk with the person whom he will eventually accuse of the crime, he starts to leave. Halfway out of the room, he stops, turns back, scratches his head, and says: "There's just this *one* thing that's bothering me . . . ." That's how it all started, with just this one thing that bothered me—a trend that went up instead of down.

During the rest of the every *Columbo* show, Falk gradually pieces the story together and figures out what really happened. And that's how this new hypothesis came about. Having noticed the mystery of the wrong-way methane trend, I wondered what might explain it and eventually found an answer in the literature of early human history that convinced me. Just about the time the methane trend began its anomalous rise, humans began to irrigate for rice in Southeast Asia. I concluded that the irrigation created unnatural wetlands that emitted methane and explained the anomaly.

That first "Columbo moment" and the subsequent investigation has been followed by other, similar mysteries: the cause of a similarly anomalous rise in atmospheric CO_2 in the last 8,000 years, the reason why new ice sheets have failed to appear in northeast Canada when the natural cycles of Earth's orbit predict that they should have, and the origin of brief drops in CO_2 that again cannot be easily explained by natural processes but that appear to correlate with the greatest pandemics in human history. But before these "Columbo moments" can be explored, we need to go back in time to see where humans came from, and to find out how and why climate has changed during our time on Earth.

PART TWO

Nature in Control

NATURE IN CONTROL

OVER MOST OF THE UNIMAGINABLY long history from our distant prehuman precursors to our more immediate and fully human ancestors, we have been passive
(or at best reactive) participants in whatever kinds of changes in climate that
nature has thrown at us. Our remote ancestors were simply too few in number,
too unsophisticated in technology, and too transiently present in any region to
leave significant footprints on the landscape or alter the climate system. Climatic
changes over these immense spans of time were entirely under nature's control.

These spans of time are beyond imagining. I suspect that even those of us who
spend our entire careers studying aspects of Earth's history have no real grasp of
their immensity. The last 2.5 million years since the time our immediate ancestors (hominids) first appeared on Earth is a tiny portion (less than 0.1%) of this
planet's entire history, yet even it is far beyond the spans of not just our individual
lives but also all written human history (covering little more than 2,000 years). As
students entering this field, we learn early on to store this information in a mental
file cabinet marked "time," with drawers labeled as billions, millions, and thousands of years ago.

In this way we file away the basic outline of Earth's story: Earth formed 4.6 billion years ago. . . . The atmosphere became oxygen-rich nearly 2.2 billion years
ago. . . . Marine organisms first left hard fossil shells after 600 million years ago. . . .
Complex life on land arrived by 400 million years ago. . . . All the continents were
mashed together into the giant supercontinent Pangaea ("all Earth") between 325
and 250 million years ago, and ice sheets formed on the parts of it near the South
Pole. . . . Pangaea broke up after 175 million years ago, and the Atlantic Ocean
came into being, while a once-larger Pacific Ocean shrank . . . No ice sheets existed
on land at the South Pole 100 million years ago in a climate that was warmer than
today. . . . India broke loose from Antarctica about 70 million years ago and began
slowly crunching into Asia 20 million years later, heaving up the Tibetan Plateau
and the Himalaya. . . . Near that time, ice appeared on Antarctica. . . . The closing
of the Isthmus of Panama joined North and South America by 4 million years ago,
and, soon after, ice-age cycles began in the Northern Hemisphere.

Later, we choose particular intervals to study, and we learn the detailed sequences of geological and climatic changes within those intervals. By degrees, and
by repetition, we become thoroughly familiar with the interval we investigate,
and eventually we become "experts." But do we really comprehend the immensity
of all that time? I doubt it.

Still, time provides us with information vital to the detective story told here. The greatest success story to emerge to date from the fourth great earth science revolution—the study of Earth's climatic history—is an understanding of the causal links between relatively small changes in Earth's orbit and relatively large changes in its climate. This important discovery arose from the convergence of knowledge from two very different disciplines: geology and astronomy. And this new understanding has two major facets: the cause of ice-age cycles that have dominated climate change at north polar latitudes, and the cause of fluctuations in the monsoons that have prevailed in the tropics.

Chapter Two

SLOW GOING FOR A FEW MILLION YEARS

SOMEWHERE IN EAST AFRICA, buried under a thin layer of soil for protection from curious eco-tourists, lie fossilized footprints 3.6 million years old. Soon after a long-ago volcanic eruption, two adult creatures walked across a bed of cooled and rain-moistened volcanic ash, leaving the marks of their feet. Occasionally protruding outside one of the two sets of footprints are extra toe marks, as if a large child was also part of the group, walking along and placing its feet inside the marks made by one of the adults, but occasionally missing the target by a little. Later, the ash hardened into rock and was buried by other sediment. Much later, erosion reexposed this deposit.

These footprints were made by creatures that were more than apes, yet certainly less than human, that walked the forests and grasslands of eastern Africa 3.6 million years ago and that are called australopithecines. The patterns in the ash show no trace of marks made by knuckles touching the ground, as would be expected if these beings used their arms as part of their natural locomotion the way modern-day apes tend to do. Fossil remains of other members of this group dating to the same or even earlier periods of time show ankle structures fully adapted for walking. Although not yet humans, these beings were clearly on a trajectory leading toward us.

Given how long ago this scenario happened, perhaps the most amazing fact about the subsequent history of our ancestors is how remarkably little things actually changed over an incredibly long period of time. Back 3.6 million years ago, our australopithecine predecessors were already acquiring their food from scavenging, gathering, and hunting small animals. Yet 3.59 million years later, just 12,000 years ago, our immediate (and by then fully human) ancestors were still subsisting by hunting, gathering, and scavenging, along with some fishing. A few tribes in remote areas live this way even now. Similarly, even though the use of stone tools began nearly 2.5 million years ago, thereby initiating the Stone Age, every human still lived in Stone Age cultures until the first use of metals just 6,000 to 5,500 years ago, not long before the early civilizations of Egypt constructed the pyramids. If you take this long view of our history, it is astonishing how close to the present most of the ancient ways of "making a living" survived.

From our highly accelerated modern perspective, it also seems amazing that our predecessors had so remarkably little to show by way of progress over spans of time stretching back millions of years. Today the term "glacial" is often used to describe the slowest imaginable pace, but somehow those sluggish ice sheets managed to grow and melt at least 50 times during the interval within which our ancestors were unable to move beyond a primitive style of existence. Almost everything in human history that we call "progress" has happened in just the last few thousand years, truly the blink of an eye.

Many narratives of human history go back to an asteroid some 10 km in diameter that slammed into Earth 65 million years ago. For the previous 100–200 million years, the largest organisms on Earth had been the dinosaurs, and our ancestor-mammals were small, rodentlike creatures trying to stay out from underfoot or scolding like squirrels from tree tops. Then, according to a hypothesis published by Luis and Walter Alvarez in 1980, the asteroid hit and wiped out the dinosaurs, or at least the larger forms we commonly think of under that name.

How this happened is still debated: terrific pressure waves from the force of the impact, instantaneous scorching from the heat generated, starvation soon afterward because all the vegetation had been burned, cooling over the next few years from an ejected plume of dust that dimmed the Sun's rays, the effects of highly acid rainwater over the next several decades, or global heating over later centuries because of the CO_2 sent from burning vegetation into the atmosphere. Despite this uncertainty, no one doubts that a major impact occurred: it left a distinctive chemical tracer (the element iridium) that is extremely rare in Earth's crust concentrated in a thin layer of sediments worldwide. And it produced gigantic tsunami waves that drove far into the coastal regions of the continents.

Over the next several years after this "impact hypothesis" was published, critics raised objections. They pointed out that many types of dinosaurs seemed to have disappeared from the geologic record well before the impact event, and they hypothesized that climatic changes that were occurring for other reasons were part or even all of the story. But this possibility has been effectively squashed by an unusual example of participatory science. In the western United States, hundreds of nonscientific volunteers were trained to hunt outcrops for fossil remains and used as brute-force labor to sieve tons of sediment in search of smaller remains such as teeth and small bones.

The results were, in a way, predictable: the more outcrops the corps of volunteers searched, and the more soil they sieved, the more bones they found. Of course. But part of what they found was revealing in a way that had not entirely been anticipated, at least not by the critics of the impact hypothesis. For every type of dinosaur that had previously seemed to go extinct at levels well below the impact layer, the volunteers found fossil remains much closer to the impact event. The closer they looked, the more their results matched the asteroid-impact hypothesis of a single and sudden extinction. These efforts implied that an infinite number of volunteers looking at every outcrop would find the actual extinctions lying right at the level of the impact event.

Because the impact killed many life forms other than just dinosaurs, it adds another dimension to the process of evolution that Darwin had envisaged more than a century ago. For tens or hundreds of millions of years, life forms apparently do compete for the available ecological niches (and means of survival) in a more-or-less Darwinian way: they outbreed or outsurvive other, similar life forms. Success in this contest is measured by small gains and losses, as if certified public accountants from insurance companies were in control of the progression of life.

But then, every hundred million years or so, a huge chunk of rock comes flying in from outer space and completely rearranges this orderly world. Most species go extinct (70% at the impact event 65 million years ago), and suddenly many once-crowded ecological niches have elbow room for the survivors, who begin to make use of it with the slower, Darwinian diversification of their talents. This side of evolution has been called "life in the interplanetary shooting gallery." Only a few sittings ducks survive the largest assaults.

As a result of the impact event 65 million years ago, the Age of Dinosaurs became the Age of Mammals. Those small, unimpressive, rodentlike mammals soon evolved into larger, more complex life forms, including the largest creatures on Earth. One line led to small animals very much like modern lemurs, tree climbers with grasping front paws and prehensile tails that lived tens of millions of years ago. That line in turn diversified and led to a primitive kind of apes that lived 10 million years ago, after which a separate group that included chimpanzees and our own ancestors branched off near 5 million years ago. By 4.5 or 4 million years ago, those australopithecines in Africa had risen up from four legs to two and were walking upright like the ones that left those footprints in the ash. The change to an upright posture can also be detected by the position of the spine at the base of the skull: for four-legged creatures, the spine enters the back of the skull; for those walking upright (including the australopithecines), the head sits right atop the spine.

The world gradually cooled during this evolution toward our species. On Antarctica, cold-adapted vegetation gave way to small mountain glaciers, and then to larger ice sheets that repeatedly grew and melted, and finally to a thick ice sheet that has stayed in place for millions of years. Around the Arctic Ocean, temperate forests gave way to cold-adapted conifers and later to tundra. Ice caps began to appear on high mountains in the tropics.

Two causes of this gradual cooling of the Earth have been proposed. Some scientists believe that the ocean is the main reason for it. The ocean carries almost as much heat poleward as the atmosphere, and its circulation is affected by plate tectonic changes, especially when continents break apart and allow the ocean to flow through a new passage, or crunch together and choke off such flows. The most frequently cited change in these tectonic "gateways" is the separation of South America and Australia from Antarctica several tens of millions of years ago. The hypothesis is that temperate ocean water that had once been diverted by land obstacles toward the South Pole and had carried large amounts of heat to Antarctica later began to flow in an uninterrupted path around the continent, leaving it isolated and cold enough to become glaciated. The other often-cited gateway change is the final closing of the Panama Isthmus some 4 million years ago. In this case, tropical Atlantic water that had once flowed westward into the Pacific Ocean through the gap between North and South America was diverted northward toward high latitudes. The hypothesis in this case is that the extra water vapor delivered northward by the warm ocean caused ice sheets to begin growing just over a million years later.

These gateway hypotheses have their doubters, and I am one. The first gateway change in the South supposedly caused glaciation by reducing the amount of ocean heat carried poleward, while the second one in the North did so by increasing it. The use of the same line of argument but in two completely opposing directions makes me doubtful about both. I also think that these gateway changes are too scattered in time and space to provide the ongoing push needed to explain the persistent drift of climate toward colder conditions over the last 55 million years or more.

A more widely accepted explanation of global cooling is a gradual drop in the amount of the greenhouse gas carbon dioxide in the atmosphere. Although CO_2 is just a tiny fraction of the total amount of gases in the air, several hundred billion tons of it are floating around up there, and the amount in the atmosphere has varied considerably through time. Think of the amount of atmospheric CO_2 as water in a partly filled bathtub. This particular tub is not very tight: it has a dripping faucet and a leaky drain, so a little water is constantly coming in and a little is going back out. The water level in the tub reflects the balance between the drippy faucet and the leaky drain.

In nature, volcanoes are the dripping faucet that add CO_2 to the atmosphere. Volcanoes are located mainly in areas where Earth's tectonic plates are colliding, especially around the rim of the Pacific Ocean. Every year, volcanoes explode somewhere on Earth, and when they do they add CO_2 to the atmosphere, like a faucet slowly dripping tiny amounts of water into a very large tub. The leaky drain for the CO_2 is right under our feet. Rainwater is slightly acidic because it contains small amounts of CO_2 from the atmosphere. Rains feed groundwater that is slightly acidic for the same reason (it contains CO_2), and the groundwater slowly weathers mineral grains in the soil. In the chemical reactions that occur during weathering, the CO_2 in the groundwater is eventually locked up in clays in the soil. In effect, it has drained out of the tub.

To cool Earth, the amount of CO_2 in the atmosphere (the level of water in the tub) has to fall. One way to do that is to reduce the amount of CO_2 coming in at the faucet. Geophysicists have shown that several kinds of plate tectonic processes that create volcanoes, such as the rate of generation and destruction of new crust at deep-ocean ridges and the creation of gigantic volcanic edifices on the sea floor, have slowed during the last 100 million years. Apparently, less CO_2 has gradually entered the atmosphere over time.

It also seems likely that more CO_2 has been going out the drain. The normally slow process of chemical weathering that occurs in soils is accelerated if the mineral particles being attacked are ground up very fine, like preparing coffee beans for dripping water in a filter. During the last 50 million years, the slow collision of India plowing northward into Asia has created the Himalaya Mountains and the Tibetan Plateau, by far the largest feature on Earth's continents. These processes create enormous amounts of ground-up rock debris. Monsoon rains that fall against these mountain slopes attack the freshly crushed rock particles and take CO_2 out of the atmosphere (down the drain) at rates much faster than normal.

The fact that both poles have cooled progressively over tens of millions of years is an argument that falling levels of atmospheric CO_2 are the primary causal mechanism. Another argument for this idea is the fact that the Antarctic continent has been centered at the South Pole for over 100 million years, yet it has had significant amounts of ice for only the last half of that interval (as CO_2 levels dropped). Only a much warmer atmosphere can explain a pole-centered continent with no ice on it.

By 2.5 million years ago, or not long after, the first members of our genus (*Homo*, meaning "man") appeared in Africa (fig. 2.1). These beings, called hominids, were somewhat shorter than us, and they had much smaller brains, about one-third the size of ours, and large brow ridges above their eyes. They fashioned crude tools by chipping and flaking pieces of stone of different hardness. Some of their food came from hunting small game and some from scavenging kills made by large carnivores. One use for the tools was to extract food: big hammer stones for battering and crushing bone to get at the marrow inside; smaller, sharper flakes for cutting through hides and scraping and cutting flesh to separate it from bone and sinew; and still other stones to mash raw meat to a hamburger-like consistency. To most of us, these tools look like random pieces of stone, but tools they were, at least in this primitive sense.

As the energy-demanding brains of our remote ancestors grew larger, they needed more protein in their diets, and meat obtained both from scavenging and from hunting small animals was a ready (although not dominant) source. These

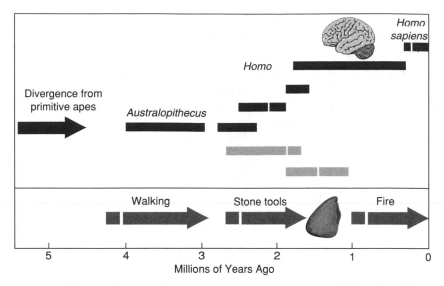

2.1. The sequence that led to modern humans during the last 4 million years includes pre-human australopithecines that walked; early members of our genus *Homo* (man) that used stone tools and controlled fire; and members of our own species (*Homo sapiens*).

protohumans lived a wide-ranging hunting-gathering life, harvesting a wide variety of nuts and berries as they became seasonally available. Some of the shaped tools were probably used to dig out tubers and roots. Like modern chimps and some primitive people, they likely used sticks to extract termites from holes and eat them. The landscape also contained many other delicacies for those with the stomach for them: bird eggs, field mice, and other fare.

As food resources were used up in one region, these hominids moved on to another. In many regions where food resources were naturally abundant, this style of life was not harsh, and only part of the day had to be devoted to satisfying basic food requirements. But the frequent movement made it difficult to accumulate large stores of food and left these beings vulnerable to extended droughts, prolonged freezes, and other extremes of weather, especially in areas less blessed with natural resources.

By about 500,000 years ago, our genus *Homo* had come to look more like us, with larger, rounder skulls holding larger brains, by then about two-thirds the size of ours. These people were members of our own species, *Homo sapiens* ("wise man"), although not yet as intelligent as modern humans. In *The Time before History*, Colin Tudge comments that these creatures "would not quite pass muster in a bus queue" today. Yet they had taken a very important step by gaining control of fire, probably at first by making opportunistic use of fires started by lightning strikes, and then later by learning how to create it at will. Control of fire brought greater protection against cold weather and wild animals, the ability to render raw meat more edible, and the possibility of manipulating the landscape by intentionally burning to improve prospects for game, in effect an early form of wildlife management.

With larger brains, these beings gradually became more adept at a range of survival skills. They could begin to draw on stored knowledge of how the game had behaved in previous years and thereby anticipate the same behavior the following year and in the future. Their improving communication skills would also have made possible increasingly clever and complex group hunting strategies. And they could use stones as crude missiles to bring down small game. These people also managed to spread out of Africa and across southern Asia. Yet, considering that 2 million years had elapsed since our genus first appeared, not much progress had been made toward modernity. Humans were still in the Stone Age, still living a hunter-gatherer life, and still making crude (but now slightly more sophisticated) stone tools.

One consequence of the highly mobile life was that children had to be spaced at relatively long intervals of four years or more. Constant relocations in search of new food resources forced people to carry all their possessions from place to place, and this required choosing between carrying infants or hauling other precious cargo that was critical to basic survival. Partly as a result of spacing their children, human populations remained small. Dependence on resources that occasionally became very scarce also helped to limit populations.

Sometime between 150,000 to 100,000 years ago, nearly modern people evolved in Africa. These people were now very much like us physically: taller than their predecessors, and with much larger brains. Like their predecessors, they still made tools by chipping and flaking stone, but they did so with increasing sophistication, producing more varied, more delicate, and better-made (though still crude) tools using more clever choices of stone. We know that they buried their dead and cared for their sick: the latter conclusion is based on finds of fossil skeletons of individuals who lived through adulthood with deformities that would have required help from others. At first, like their predecessors, they continued to rely mainly on smaller game that could be killed with little or no danger to the hunter.

It is odd to think about highly intelligent human beings living such incredibly primitive lives. Their brains were capable of everything ours are now, but they lacked the base of common knowledge available today. If one of their babies were somehow brought into the modern era and raised in today's society, he or she would have just as good a chance as any of us to become an astrophysicist, a carpenter, or a billionaire manufacturer of malfunctioning computer software. This

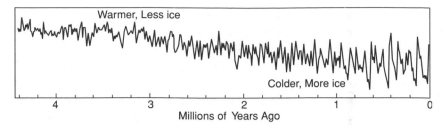

2.2. Over the last 5 million years, as modern humans evolved, Earth's climate slowly cooled.

juxtaposition may seem odd, but much the same thing has happened in recent decades when tribes that happened to have escaped the onrush of civilization were abruptly exposed to its complexities. People from many primitive cultures quickly adapted to modern ways.

This long story of prehuman and early human evolution played out in a world that continued to grow colder (fig. 2.2). Nearly 2.75 million years ago, the cooling reached a critical threshold, and ice sheets first appeared in high latitudes of the Northern Hemisphere. This ice was not permanent; it grew and melted as changes in Earth's orbit delivered varying amounts of sunlight to north polar regions. Through time, the glacial cycles grew larger.

Through these millions of years, our ancestors had no lasting effects on the environment or on climate. At some point they began using "fire sticks" to burn grasslands and thereby altered their local environment temporarily. But dry-season lightning strikes had been doing the same thing for the hundreds of millions of years that vegetation had been growing on land. Burning leaves the roots of the grasses untouched, while it fertilizes the soil and causes grass to grow back even more prolifically. Burning also releases CO_2 to the atmosphere, but new grass reclaims it through photosynthesis during the next growing season. The people move on, and they leave no permanent trace of their presence on the landscape. Nor do they affect the level of water in the tub.

Chapter Three

LINKING EARTH'S ORBIT TO ITS CLIMATE

THE REALIZATION THAT small changes in Earth's orbit might have regular and predictable affects on climate originated just over a century and a half ago. It came about from a convergence of knowledge in two very different disciplines: the still-young science of geology and the somewhat older field of astronomy.

In the middle 1800s geologists first proposed that great, mile-high ice sheets had existed in large areas of the Northern Hemisphere and had disappeared in the not-too-distant past. These great masses of ice were called "sheets" because of their dimensions: thousands of miles in length and width and a mile or two in thickness, or just about the same relative proportions as the sheets on your bed.

The primary evidence for such vast covers of ice was the presence of long, curving ridges of rubble running for miles across the landscapes of northern North America and northern Europe. These ridges of jumbled debris, called moraines, contain everything from fine clay to sand, pebbles, cobbles, and boulders. For a long time, most scientists had believed them to be the result of the great biblical flood, which they imagined would have had the power to move even the largest debris across the landscape. The fresh, young-looking appearance of these ridges of rubble also seemed consistent with a biblically derived estimate that Earth was just under 6,000 years old.

But water tends to sort the material it carries by size, leaving deposits rich in sand and silt and clay in separate regions. In contrast, these piles of rubble were just that: everything from clay to boulders all heaped up together. Naturalists like Jean de Charpentier who lived and worked near the Alps and the mountains in Norway had noticed the same kinds of rubbly deposits lying at the edges of active mountain glaciers in mountain valleys. Ice is nature's messiest housekeeper: it carries and pushes all sizes of sediment and dumps it in great heaps wherever it melts. Also found in northern and montane regions were large boulders called erratics that were totally unlike the underlying bedrock. It seemed obvious that these boulders had been plucked from bedrock in far-away areas, carried long distances, and dumped onto the landscape.

By the late 1830s Charpentier and others had convinced Swiss geologist Louis Agassiz that the deposits in the alpine valleys were created by mountain glaciers, as were the gouges and scars left where glaciers had ground loose rocks across the underlying bedrock. But Agassiz took a much larger step: he proposed that ice also accounted for the other bands of jumbled debris and the bedrock scars found all across northern North America and northern Europe. He claimed that great ice sheets had once covered large parts of these continents. At a time when few, if any, were even aware of the existence of the Greenland ice sheet and the massive Antarctic ice sheet had not yet been discovered, this notion seemed absurd.

Agassiz borrowed a term previously used by botanist Karl Schimper to describe this cold interval as an "ice age." Energetic and persuasive, he began a long campaign to convince the scientific community that these enormous sheets of ice had indeed existed. Although his challenge to the scientific wisdom of the time was widely resisted at first, little by little other prominent scientists went out into the field, looked at the evidence, and concluded that Agassiz must be basically right. By the 1860s Agassiz's campaign had largely succeeded.

But this important discovery immediately raised a troubling question. Even though the technique of radiocarbon dating would not be discovered until the 1940s nor widely used until the 1950s, geologists could tell simply from the fresh look of the sediments that the glacial deposits were not very old. Most rock deposits from Earth's long history are hard and compact, and they look ancient, even at a glance. These piles of debris were loose, crumbly, and much fresher looking, and therefore obviously much younger. One way to estimate the time since the glaciers had melted was to examine sediments in lakes within or near glaciated areas and count the number of annual layers called varves, which are alternating light-dark bands deposited each year. Records spliced together from sediments in many Scandinavian lakes suggested that the ice had melted between 12,000 and 6,000 years ago, an estimate that turned out not to be far off.

The fact that such enormous masses of ice had existed so recently posed a major challenge to geologists who were only beginning to accept the theory devised by James Hutton in the 1700s that Earth's surface features result from geologic processes that work over long intervals of time. If it really takes tens or hundreds of millions of years to build or erode a mountain range, how could such processes create or destroy ice sheets within just thousands of years and, in the case of the recent ice melt-back, within just the last few thousand years?

Meanwhile, other features that also looked very "young" and thus seemed to indicate recent changes in climate were being discovered elsewhere. In the late 1800s American explorers on horseback came into the Great Basin desert region of the Southwest. Throughout the region they found features that clearly marked the shorelines of former lakes that had once flooded these basins to levels well above those of modern lakes or basin floors. Surrounding modern-day Salt Lake City, geologist Grove Gilbert found fresh-looking beaches up to 300 meters (more than 900 feet) above the level of today's Great Salt Lake. These old shorelines were marked by notches cut into rock outcrops by wave action and by fresh-looking lakeshore sediments deposited in continuous, flat-lying terraces visible from miles away. The obvious implication was that an enormous lake must once have flooded the entire basin between the surrounding mountain ranges, covering some 50,000 square kilometers (20,000 square miles), almost the size of modern-day Lake Michigan, and inundating the site of present-day Salt Lake City. Smaller lakes, also much larger than those present today, were scattered across the basins of the American Southwest. This entire region, not very long ago, had once been much wetter.

The earliest geologists and geographers who penetrated the interior of North Africa were finding similar evidence in and south of the Sahara Desert. Surrounding several basins where modern-day lakes are small or dried up entirely are the same kinds of wave-cut notches and sediment terraces, indicating that larger and deeper lakes must have existed in the recent past. Scattered remains of these older lake-bed sediments can be found in regions where the fierce Saharan winds have not yet blown them away. The full catalogue of young-looking deposits of this kind from around the world is long, but all of the examples point to the same basic conclusion: something has been causing enormous changes in climate on relatively short time scales, and that something cannot possibly be processes involved in Earth's slow tectonic changes.

Still another complication emerged. During the 1800s and early 1900s, it gradually became clear that large ice sheets and higher lake levels had existed more than once in the relatively recent past. In a few northern locations, the glacial debris heaped in moraine ridges lay on top of a layer of soil very much like the soils that typically accumulate today in temperate regions. Pollen grains found in these soils came from heat-adapted trees like oaks and hickories. This evidence indicated that a warmer (nonglacial) climate much like that today had existed before the ice sheet arrived and deposited the rubble. But then, lying underneath the older soil, scientists found another layer of glacial debris indicating another earlier glaciation in the same region. Explorers in now-arid regions also found evidence of more than one earlier interval of higher lake levels.

These discoveries meant not only that climate had been colder or wetter in the very recent past, but that it had switched back and forth more than once between the warm and cold or wet and dry states. A few geologists made brief but futile attempts to explain this evidence by invoking fluctuations in Earth's crust over intervals of a few tens of thousands of years, but these explanations convinced few.

As it turned out, the answer to this mystery was to be found not in Earth's internal geologic processes, but in its orbit in space. Centuries earlier, during the 1500s and 1600s, astronomers Nicholas Copernicus, Johannes Kepler, and Galileo Galilei had discovered that Earth is not the center of the universe but a small planet held in an orbit around the Sun by the pull of gravity across 93 million miles of space. Similarly, Earth's gravity field holds the Moon in orbit some 242,000 miles away, just as it holds us securely right here on terra firma.

In time, other astronomers began to investigate another, more subtle effect of gravity—the influences on Earth of the combined gravitational tugs of the Sun and Moon and all the planets as Earth orbits the Sun. Gradually they had come to learn that the secondary tugs of the planets, even though much smaller than the pull of the Sun, have important effects. Jupiter, 400 million miles away, is large enough to pull noticeably at Earth's orbit. Our moon, although relatively small, is close enough to tug on Earth. These gravitational pulls are aided by the fact that Earth's shape is not perfectly spherical but bulges out slightly at the equator, forming an irregularity on which the external gravitational attractions can act.

In 1842, just a few years after Agassiz's dramatic claim that ice sheets had once covered large parts of the Northern Hemisphere, astronomer Joseph Adhemar made a wonderfully imaginative mental leap that would eventually lead to a theory of the ice ages. Adhemar knew that some aspects of Earth's orbit had changed over intervals as short as a few tens of thousands of years. His brilliant "eureka" realization was this: changes in Earth's orbit should affect the amount of solar radiation reaching its surface, which in turn should have an impact on climate, including the appearance and disappearance of ice sheets. Although this wonderful insight proved correct, it was only a beginning, a conceptual basis on which to build by finding the actual mechanisms that link changes in Earth's orbit to changes in its climate.

One way to alter the Sun's effect on Earth's climate is to change its height in the sky. Summers today are warmer than winters in part because the Sun is higher in the daytime sky and provides more warmth, and winters are colder because the Sun is lower and its weak radiation provides less heat. By direct extension, anything that affects the Sun's height above the horizon over much longer time scales should affect long-term climate.

Of course, even though our Earthbound perspective makes it look as if the Sun is changing, in fact that is not the case. The real reason for the apparent changes in the Sun's elevation lies in Earth's orbit. Earth's rotational axis is tilted relative to its orbital path around the Sun. The angle of tilt remains constant at 23.5° throughout the yearly orbit, as does the direction in space toward which Earth is tilting or leaning. In summer, Earth is in the part of its orbit where it is tilted directly toward the Sun, and so the Sun appears high in the daytime sky. But by winter, Earth's orbit has reached the opposite extreme, and its tilt is now directed away from the Sun, so the Sun appears lower in the sky. So the height of the Sun in the sky, and the amount of solar radiation it delivers, actually results from the interplay between the Earth's axial tilt and its position in its annual orbit. Those sixteenth-century astronomers already had this figured out.

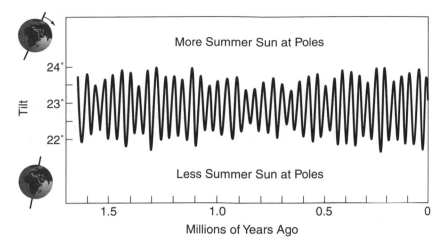

3.1. Gradual changes in the tilt of Earth's axis relative to its orbit around the Sun at a cycle of 41,000 years alter the amount of solar radiation that arrives at high latitudes.

Over long time scales, Earth's angle of tilt does not remain constant. In the 1840s French astronomer Urbain Leverrier found that the gravitational attraction of large planets (mainly Jupiter) causes Earth's tilt to vary within a range of 22.2° and 24.5° over a cycle that is 41,000 years long. Every 41,000 years, the tilt goes from a maximum to a minimum and back to a maximum. These tilt cycles are regular in both length and amplitude (fig. 3.1).

These slow variations of about 2.3° in the angle of Earth's tilt alter the height of the Sun in the daytime sky. Although 2.3° sounds like a trivial change, it still matters at high latitudes where the Sun is always very low in the sky. North of the Arctic Circle and south of the Antarctic Circle, the Sun never rises in midwinter (the endless "polar nights"). In midsummer the Sun never sets but instead drifts in a slow circle around the horizon at a very low angle. Even a small shift in the Sun's low elevation makes a measurable difference in the amount of solar radiation it delivers.

You can appreciate the importance of small changes in Sun angle by a familiar example from middle latitudes: gently sloping hillsides facing south receive enough solar radiation to melt a thin layer of snow quickly, while the north-facing slopes of those same hills may remain snow-covered for several days longer. The small difference in the angle of the Sun's rays against the slopes accounts for the different rates of snow melt. Similarly, small changes in Earth's tilt over its 41,000-year orbital cycle can cause significant differences in solar radiation at latitudes poleward of 45°.

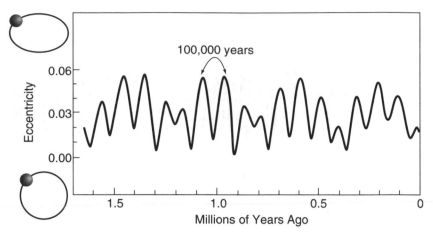

3.2. Gradual changes in the eccentricity ("out-of-roundness") of Earth's orbit around the Sun occur at a cycle of 100,000 years.

The second way to alter climate is by changing Earth's distance from the Sun. You can make this claim somewhat tangible by holding your hand a foot away from a typical light bulb and then holding it 3 inches away. Distance from a heat source clearly matters, and the amount of heat from a light bulb is in the general ballpark of the average amount coming from the Sun.

Two factors in Earth's present-day orbit combine to alter Earth's distance from the Sun during its annual revolution. The first is the eccentricity of its orbit. Although we generally think of the orbit as circular, in fact it is slightly out of round, or elliptical, in shape. As a result, Earth is about 5 million kilometers (3 million miles) closer to the Sun in one part of its orbital path than on the opposite side. These changes are small but significant departures from Earth's average distance to the Sun of 155 million kilometers (93 million miles).

Over long time scales, this elipticity, commonly called eccentricity, changes (fig. 3.2). French astronomer Leverrier is again credited with discovering that eccentricity varies at a cycle near 100,000 years. On rare occasions the eccentricity drops to zero and Earth's orbit around the Sun becomes perfectly circular. Most of the time the orbit is eccentric, with the amount of eccentricity constantly varying. These changes in eccentricity are more irregular than those of tilt: the peaks and valleys vary widely in size. Changes in eccentricity through time affect the Earth-Sun distance in different parts of Earth's orbit because they produce departures from circularity.

The second aspect of Earth's orbit that affects Earth-Sun distance is precession, a wobbling motion best understood by analogy to a spinning top. Like a top,

Earth spins (rotates) once per day on its tilted axis. Earth revolves around the Sun once a year, a motion also shown by most tops that slowly trace out a circular path on a flat surface. But tops often show a third kind of motion: they may wobble or change the direction in which they lean, first toward one direction, then another. These changes in the direction in which Earth leans on its tilted axis differ from the amount of lean, which is the tilt.

In the 1800s French mathematician Jean le Rond d'Alembert was the first to understand how precession affects Earth's orbit around the Sun. He found that it takes about 22,000 years for Earth's tilted axis to complete one slow wobble in its orbit, an interval far longer than its once-daily rotational spin or its once-yearly revolution. It takes 22,000 annual revolutions and over 8 million daily rotational spins for Earth to complete just one of these wobbles in its orbit. You would have to watch a top carefully for a long time to detect so slow a wobble.

At the start of a single precession cycle, Earth is tilted in a particular direction in space, and it keeps that same attitude throughout the year. But as the years pass, the direction of tilt slowly begins to shift, gradually tracing out a circle. After 11,000 years, Earth's tilt direction has shifted enough that it leans in exactly the opposite direction it did initially. Then, after another 11,000 years pass, the direction of tilt circles all the way around to the same position it had at the start, 22,000 years earlier. In trying to explain precession in the classroom, I usually ended up walking around in a circle with my body and arms tilting in one direction and then in the other, apparently doing some religious dance ritual (well, of course, a Sun dance!). Most of my students probably remembered my weird body English long after they forgot how precession works.

Eccentricity and precession work together to determine the amount of solar radiation actually arriving on Earth. In effect, changes in eccentricity (fig. 3.2) act as a multiplier on the cycles of precession. They make the precession cycles larger in amplitude when eccentricity is high and smaller when eccentricity is low. Each individual precession cycle stays nearly 22,000 years, but the multiplier effect from eccentricity determines whether the swings from peaks to valleys are large or small (fig. 3.3).

If Earth's orbit around the Sun were perfectly circular (with zero eccentricity), and if Earth did not slowly wobble (precess) in its orbit, the amount of solar radiation received every summer would be identical, and every winter as well. More radiation would arrive in summer than winter because of Earth's tilted position relative to the Sun, but the amounts of radiation received during each of the seasons would not change from millennium to millennium.

But because Earth's orbit is eccentric and also precesses, summers and winters do not remain identical through time. When Earth's orbit is highly eccentric, it can be as much as 6 percent closer to the Sun during part of its annual orbit and

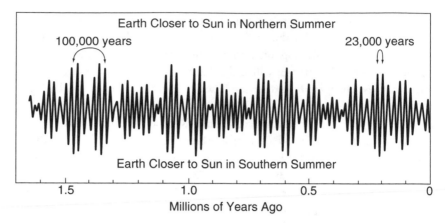

3.3. The combined effect of changes in the eccentricity of Earth's orbit and in its slow wobbling motion cause changes in the amount of solar radiation received at low and middle latitudes at a 22,000-year cycle, with the size of the cycles varying at 100,000 years.

6 percent farther away at the other extreme compared to a perfectly circular orbit. Precession makes these inequalities in Earth-Sun distance shift slowly around the eccentric orbit so that over thousands of years they occur during different seasons. As a result, the solar radiation received during each season varies through time.

Fortunately, all of these complicated interactions of eccentricity and precession are neatly summarized in the single curve shown in figure 3.3. The effects of this signal on solar radiation are felt at all latitudes, except for regions lying in the darkness of the winter polar night. So, in the end, it all comes down to two wiggly lines. Changes in tilt at a cycle of 41,000 years affect solar radiation at high latitudes (fig. 3.1). And all latitudes on Earth are affected by the 22,000-year cycle of precession, with its eccentricity multiplier effect at 100,000 years (fig. 3.3).

Because these changes in Earth's orbit are relatively small, they might seem unlikely to have much effect on the amount of solar radiation received at any location, but this is not so. For any particular season of the year and location, changes in solar radiation can vary by more than 10% around the average value. This range of variation is larger than the present-day difference in radiation between Quebec and Atlanta, or between London and Lisbon.

For decades many scientists resisted the hypothesis of orbital control of climate. One counterargument was that changes in solar radiation at any location on Earth in any particular season are invariably opposed by the opposite sense of change in the opposite season. If extra radiation arrives in summer, then winters at that same spot will have a deficit of about the same size. The same thing is true for the hemispheres. When Northern Hemisphere summers are receiving extra radiation, Southern Hemisphere summers are receiving a deficit of about the same size. This opposing sense of seasonal and hemispheric radiation trends results from the way our planet moves between the opposite extremes in its orbit through the year. If Earth is unusually close to the Sun in one season, it will inevitably be unusually far away during the opposite season on the other side of its eccentric orbit. And if Earth leans toward the Sun in one season, it must lean away from the Sun six months later, during the opposite season. Over the course of a full year, and over the entire planet, these changes cancel out and leave no significant difference in the amount of radiation received by the planet as a whole. So, for seemingly valid reasons, the critics concluded early on that Earth's net response to long-term differences in solar radiation would be trivial.

But these early critics overlooked something important about the climate system. They were thinking of a homogeneous Earth that reacts in exactly the same way everywhere. This viewpoint made no allowance for the complexity of the climate system, or for the different ways its many components might respond. An obvious example is the distribution of land and ocean. Most of the Southern Hemisphere is ocean, and oceans tend to moderate, rather than amplify, changes in climate. Large amounts of incoming solar radiation are stored in the oceans, but in a layer thick enough that the changes in temperature right at the surface of the ocean are actually modest. In contrast, the Northern Hemisphere has the world's largest continents (Eurasia, North America, and North Africa), and land surfaces are far more reactive to climatic pushes than the ocean. When solar radiation heats land surfaces, the heat can penetrate only a short distance into the sediment or rock, so the land heats up more quickly, and to a far larger extent, than the ocean.

These fundamentally different behaviors are evident in the relative response of oceans and land masses to the changes in solar heating during our present-day seasons. Maximum heating in the Northern Hemisphere occurs at the summer solstice (June 21), and the interiors of the northern continents reach maximum temperatures a month later, during July, yet the Northern Hemisphere oceans do not reach their seasonal maxima until late August or early September. In addition, the land in the center of large continents seasonally warms up 5 to 10 times as much as do the oceans at the same latitude. In winter the land also cools off faster than the ocean, and by a far larger amount. For these reasons, the land-rich Northern Hemisphere does not respond in the same way as the ocean-dominated Southern Hemisphere. The critics failed to allow for differences like these.

Earth's surfaces also vary widely in the amount of flat terrain versus mountains, deserts versus forests, and sea ice versus open (ice-free) water. These differences in surface characteristics also produce different regional responses to changes in solar radiation. As we will see in the next two chapters, the components of the climate system that matter most in a particular region may react more strongly to the extra solar radiation arriving during one season than to the deficit of radiation that arrives during the opposite season. In such cases, the opposing seasonal radiation trends are not canceled out in the net response of the climate system. The critics of the concept that orbital changes could drive climate also failed to allow for this possibility.

Meanwhile, ignoring the critics, a few scientists were working at the task of trying to determine the specific mechanisms by which changes in Earth's orbit might drive changes in its climate. In 1875 James Croll of Scotland made preliminary calculations of changes in the amount of solar radiation received on Earth. In 1904 German mathematician Ludwig Pilgrim published a paper summarizing a decade of tedious work calculating the variations of Earth's orbit over the last three million years. Although Pilgrim made no attempt to tie these calculations to climate changes, his work was essential to efforts that followed.

Beginning in 1911, astronomer Milutin Milankovitch made a series of laborious hand calculations of the amount of solar radiation received by latitude and by season over the entire Earth during the last several hundred thousand years. His calculations continued in a jail cell where he was imprisoned by the Austrians during World War I and then afterward when he was paroled. He took into account the two factors that Isaac Newton had centuries ago shown to be the major controls on solar radiation: (1) the varying angle of incoming solar radiation relative to the surface of the Earth (the effect of tilt), and (2) Earth's distance from the Sun (the combined effects of eccentricity and precession). These laborious calculations, now done with much greater accuracy by computers in just minutes, laid the groundwork for many discoveries, the most important of which are covered in the next two chapters.

Chapter Four

ORBITAL CHANGES CONTROL ICE-AGE CYCLES

IMAGINE STANDING ON a tropical island in the Pacific, one of those coral atolls that barely rises high enough above sea level to avoid being inundated by passing typhoons. Now imagine standing in that exact same spot 20,000 years ago. What would be different? The waves would still be crashing against the edges of that same island, but now some 375 feet below the place you stand. The difference would be the result of water taken from the ocean and stored in ice sheets.

After almost 40 years' study of past climates, nothing amazes me more than the fact that enough water was stored in the great ice sheets of North America and Europe during the last glacial maximum 20,000 years ago to lower the level of the entire world ocean by that amount. The next time you stand at the coast looking out to sea, knowing you are looking at just a tiny part of that one ocean, try to imagine draining the world ocean by 375 feet. Or the next time you fly across the Atlantic or Pacific, taking hours to get to the other side at 400 or 500 miles per hour, try to imagine.

With all that water withdrawn from the ocean, you could walk from Ireland to Britain and then on to the mainland of Europe, or from Tasmania to Australia to New Guinea, or from Siberia to Alaska, or from the mainland of Southeast Asia through Indonesia all the way to Borneo. The modern-day Sea of Japan was then just a coastal lake. The Persian Gulf was dry land. In flatter coastal regions, the coastline was tens or even hundred of miles seaward of its present location.

Or consider how much ocean water is trapped in the ice on land. Viewed from space today, the largest feature in North America is the broad bulge of bedrock extending from the high plains of Colorado and Wyoming westward across the Rockies and the Colorado Plateau to the Sierra, Cascade, and Olympic Mountains of the far West and Pacific coast. On average, this terrain bulges about 2,500 meters (8,000 feet) above sea level. Yet, 20,000 years ago, the great ice sheet of North America formed a bulge this same size. At its crest, the ice sheet was at least 2,000 meters (6,500 feet) high, and broad, rounded domes of ice covered all of Canada and parts of the northern United States. The weight of all this ice caused the bedrock underneath to sag at least 600 meters (2,000 feet) below its present-day level, forming a bowl-shaped depression. The mass of the ice was so large that Earth's axis of rotation began to shift slowly toward the ice center.

This great feat of natural engineering, dwarfing anything humans have done, has astounded generations of scientists, and the resulting sense of awe has inspired some of them to life-long pursuit of explaining how this could have happened. For those who sensed that orbital changes might be the explanation, the first problem was to figure out which season was critical and why.

In the 1860s and 1870s, James Croll proposed that the amount of solar radiation received in winter is the critical control on the size of ice sheets. At first, this idea sounds entirely sensible: snow must fall in order for ice to accumulate, and winter is certainly the most glacial-like of the seasons. Lower amounts of solar radiation would make winters colder and longer, allowing more snow to fall, and favoring the growth of ice sheets. But on further reflection, this idea makes little sense. Croll had overlooked the power of ablation, a word that summarizes all of the processes that melt snow and ice. Had Croll lived through the warm summers here in the southern Shenandoah Valley of Virginia, he might have seen things differently. Try to imagine even the thickest snow bank in some deep mountain cove surviving the blazing Sun, midday heat, and warm rains of summer.

Even the winters around here don't let snows lie on the ground for long. In early January 1996, we had an unusual 30-inch blizzard of powdery driven snow that drifted to depths of 5 and 6 feet. This storm hit the entire east coast from Georgia to Maine and was one of the biggest storms of the 1900s. After the snow stopped, the temperature stayed cold for about a week, as the powder slowly settled to 17 inches of more compacted snow.

One evening the local forecast called for rising temperatures and rain overnight. Not long before midnight, and before the rain started, I turned on our floodlights, looked out at the deep snow, and went to bed. That night, strong rains pounded on the roof of our house, bringing me half-awake several times. Around 6 in the morning, I got out of bed and turned on the floodlights. To my disbelief, the meadow was almost completely free of snow, except for a few patches where the drifts had been deepest. Somehow, almost a foot and a half of compacted snow had disappeared in less than 6 hours. I later learned that the temperatures overnight had risen into the 50s and 60s (°F). The rain that fell was so warm that it worked on the snow like warm faucet water on an ice cube. That night, the local hillsides shed floodwaters normally seen only during heavy summer cloudbursts. Here was an example of rapid ablation even during the night and in the middle of winter! At this latitude (37°N), snow rarely covers the ground for more than a month per year each winter because of the effect of the Sun and above-freezing daytime temperatures.

To our north, of course, ablation is not so dominant. Across the populated southern tier of Canada (near 50°N), snow may lie on the ground for as much as half the year. Farther north, at latitudes above 65°N, the snow-free season shrinks to just two months. Yet in today's climate, no matter how snowy a winter may be, the summers are still warm enough to melt all the snow that falls, except for scattered patches during unusually cool summers. Even that far north, it is impossible to leave any significant amount of snow in place at summer's end so that the next winter's accumulation can add to it and start the process of building an ice sheet. This key insight—the power of summer melting—was missing from James Croll's hypothesis.

In the early 1900s Milutin Milankovitch, following up on guidance from meteorologist Wladimir Koppen, proposed that the amount of solar radiation received in summer is the critical factor that determines the growth and melting of ice sheets. He hypothesized that ice sheets would have grown during times when summer radiation was low and melted when it was high. In those far northern regions of the Arctic where snow barely melts in modern summers, a small decrease in solar radiation would allow some snow to persist through the summer in favorable locations. With more snow added the next winter, the snowfields would expand, and their bright surfaces would reflect more radiation and chill the climate further, allowing still more snow and ice to accumulate.

Milankovitch's idea that summer is the critical season initially made some headway, but most climate scientists were not convinced that he was correct. Still lacking was persuasive supporting evidence based on Earth's actual climatic history. For half a century, Milankovitch's idea remained in limbo, neither completely supported nor rejected by the evidence. He died in 1958, just as the first critical evidence testing his hypothesis began to emerge.

The first important advance was the development of radiocarbon dating by Willard Libby and colleagues in the late 1940s. Climate scientists finally had a means of dating some of the deposits left by the ice sheets. Radiocarbon dating is normally useful for ages back to about 30,000 years ago, so the piles of glacial debris (moraines) strewn across the landscapes of northern North America and Europe could be dated if they contained organic carbon, or their age could be constrained by dating the carbon-rich soil layers lying immediately above or below the rubble. Most of the glacial deposits yielded ages of 20,000 years or younger.

Other kinds of deposits confirmed this finding. Scientists working on cores taken from lakes in areas south of the former ice sheets found that pollen grains preserved in the deeper lake sediments had come from trees that were far more cold-adapted than the forests growing in those regions today (for example, spruce pollen grains in regions where oak forests grow now). Radiocarbon dating confirmed that the layers of pollen from the cold-adapted trees were the same age as the glacial debris.

Radiocarbon dating also showed that the large ice sheets of the last glacial maximum had gradually melted between about 16,000 and 6,000 years ago, consistent

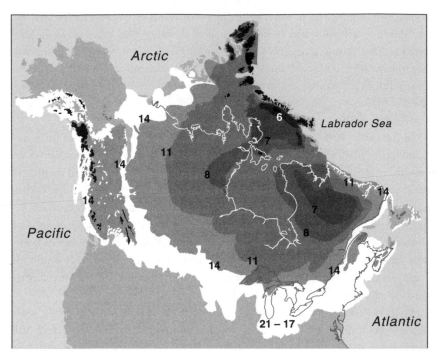

Arctic

14

6   Labrador Sea

11

14   11   8   7   11   14

14   7

Pacific   8

14   11   14

21 – 17   Atlantic

4.1. The last major ice sheet on North America slowly retreated from the northern United States into far northeastern Canada between 20,000 and 6,000 years ago. Numbers show past ice limits in thousands of years.

with Milankovitch's theory. During that interval, solar radiation in Northern Hemisphere summers was as much as 9% higher than it is today because Earth's poles were tilted more toward the Sun, and because Earth was closer to the Sun during the northern summer. Over those 10,000 years, the ice sheets slowly retreated, ending up as small remnants in northeast Canada (fig. 4.1).

Radiocarbon dating also showed that the ice sheets had earlier reached glacial-maximum size a few thousand years after an interval when solar radiation had been considerably lower during Northern Hemisphere summers. This lag of ice-sheet size behind solar radiation had also been predicted by Milankovitch, who had noted that ice sheets do not grow or melt immediately but take thousands of years to respond to this solar "driver."

But a single glaciation could hardly be taken as a definitive test of the Milankovitch hypothesis. Scientists needed a long record spanning many glaciations to compare with the solar radiation changes calculated from astronomy. Unfortunately, this record was not to be found in the regions where the ice sheets had existed. Ice sheets are agents of erosion that bulldoze loose material lying in their paths. Each successive glaciation destroys most of the rubble deposited by previous ice advances. The only record left largely intact is the retreat of the ice sheet from its most recent maximum 20,000 years ago (fig. 4.1). Think of a classroom blackboard used for a day of lectures. With each erasing, the record of earlier lectures is lost, except for scattered words around the edges. In a similar way, only a few fragments of previous glaciations remain untouched by later erosion, and those are usually too old to be dated by the radiocarbon method. So the place where the ice sheets existed is not the place to look for a continuous record of their history.

In contrast, ocean basins are places where deposition dominates. Shallow coastal regions are subject to wave erosion, but most deep ocean basins lie beyond the reach of even the largest storms. For this reason, in the middle and late 1900s, attention turned to ocean sediments as a possible archive of the complete history of the ice ages. Techniques were developed for dropping large, hollow pipes into the soft, sandy mud of the sea floor during oceanographic expeditions to retrieve sediment cores. In many places, these cores provided continuous records extending hundreds of thousands of years into the past.

Cores retrieved from ocean basins adjacent to the ice sheets were found to contain layers of debris delivered directly by the ice—fragments of continental rocks like granite and smaller mineral grains like quartz and feldspar. Initially, this debris had been incorporated into the ice sheets as they eroded the land. Then, as ice flowed out from the continental interiors and reached the sea, icebergs broke off, floated away, melted, and dropped the debris into the sediments on the sea floor far from its source regions. In contrast, during intervals when the ice sheets disappeared, ocean sediments were free of continental debris. These layered sediment sequences from the ocean held long and complete records of the presence and absence of ice sheets.

Ocean sediments also contain an even more important measure of the size of the ice sheets. Like other elements found on Earth, oxygen comes in several forms called isotopes, which differ slightly in mass and weight. Both ocean water and ice (liquid and solid H_2O) contain oxygen-18 and oxygen-16, in amounts that vary from region to region. When ice sheets form, they take more of the lighter oxygen-16 isotope from the ocean and store it as ice, leaving more of the heavier oxygen-18 isotope behind in the ocean. Living in the ocean are hard-shelled plankton that form $CaCO_3$ shells, and the oxygen in their shells comes from seawater. When the plankton die, their shells fall to the sea floor and become part of the permanent record piling up over time.

The layered history of deposited plankton shells tells us about the ice sheets. When the ice sheets grow, and more oxygen-16 is taken from the ocean, the shells deposited in the ocean sediments are enriched in oxygen-18. When the ice sheets melt, the extra oxygen-16 is returned to the ocean, and the plankton living at that time incorporate it in their shells. By analyzing layer after layer of fossil shells in sediment cores, scientists can reconstruct the history of growing and melting ice sheets even in regions far from the ice.

Pioneering investigations of sediment cores by geochemist Cesare Emiliani during the 1950s and early 1960s had shown that a sequence of at least 5 glacial-interglacial alternations had occurred over an interval thought to span the last several hundred thousand years. The fact that these glacial-interglacial repetitions occurred in rather regular cycles gave a major boost to the Milankovitch hypothesis because Earth's orbit also changes in regular cycles (chapter 3). But the age of these oscillations was still uncertain because these records extended far beyond the range of radiocarbon dating.

The problem of dating these younger cycles was soon overcome by a clever but independent approach that made use of fossil coral reefs as sea-level "dipsticks." Coral reefs form at or just below sea level and leave hard skeletal structures in the fossil record. As ice sheets grow, they take water from the ocean, and when they melt they return it. As the level of the ocean moves up and down the sides of ocean islands, the living coral reefs follow, since they can grow only near sea level. Over time, the fossilized corals leave behind a history of past sea level along the island slopes, and changes in sea level can be used to calculate the volume of seawater trapped in the ice sheets. Also coming into use during the 1960s was a method for dating coral reefs by measuring small amounts of radioactive uranium that exists in seawater and is incorporated in the skeletons of corals. Uranium allowed the corals to be dated.

Three coral reefs that formed between 80,000 and 125,000 years ago proved to be invaluable in dating the last interglaciation, a time, like today, of minimum ice volume and high sea level. The dates of these reefs matched three levels in the ocean cores where the oxygen isotopes and the scarcity of glacial debris indicated minimum amounts of glacial ice. These reefs helped to anchor the time scale of the ice-age cycles and set up a major discovery.

In 1976 marine geologists Jim Hays and John Imbrie and geophysicist Nick Shackleton published the first strong confirmation that ice-age cycles are tied to variations in Earth's orbit. Using a marine sediment record extending back more than 300,000 years, they found three major cycles of variation in the oxygen-isotope (ice-volume) record: cycles of 100,000, 41,000, and 22,000 years, precisely equivalent to the orbital cycles of eccentricity, tilt, and precession (chapter 3). In addition, the timing of the ice-volume cycles at 41,000 and 22,000 years matched the prediction made by Milankovitch: it lagged several thousand years behind the changes in Northern Hemisphere solar radiation. The story of this exciting discovery, and related efforts, is told in a book written by John Imbrie and his daughter Katherine, *Ice Ages: Solving the Mystery*.

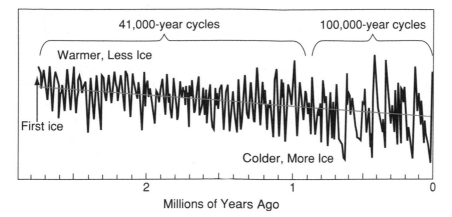

4.2. Large ice sheets first appeared in the Northern Hemisphere nearly 2.75 million years ago and grew and melted at the 41,000-year cycle of orbital tilt until about 0.9 million years ago. Since that time, the major cycle of ice-sheet changes has been at a cycle of 100,000 years.

By the 1980s, a ship funded by an international scientific consortium had begun applying techniques borrowed from the oil industry to drill sediment sequences penetrating thousands of feet into the sea floor and millions of years back in time. From these long sequences came the first complete record of the history of glaciation in the Northern Hemisphere. Several methods were used to date these much longer sequences. One technique relied on magnetic signatures carried by tiny, iron-rich minerals present both in ocean-sediment layers and in layers of basaltic rock formed from cooling lava. The magnetic patterns were initially found in the rock layers and were dated using radioactive decay methods. Then the dating scheme from the rocks on land was directly transferred to similar patterns found in the ocean sediments, providing a time scale for the glacial variations.

Scientists at last had the full history of Northern Hemisphere ice-age cycles shown in figure 4.2. One of the surprises about this record was the number of glacial cycles: at least 40–50 of them, depending on how many of the smaller cycles embedded in larger ones you choose to count. The fragmentary glacial record left on the continental "blackboard" after all those erasings had suggested 4 or 5 glacial cycles. Now it was clear that 10 times that many had occurred.

These records revealed the full history of the northern ice ages. In the Southern Hemisphere, a large ice sheet had been present for at least 14 million years on the pole-centered Antarctic continent. In contrast, the North Pole is located in the Arctic Ocean, and the nearest continents on which ice sheets can form lie at lower latitudes, where the summer Sun is stronger. For many millions of years, it was just too warm for ice to form in the North, even when the orbital changes caused minima in solar radiation. As a result, no ice sheets existed in the Northern Hemisphere 3 million years ago.

But global climate was steadily cooling (chapter 2). The first large ice sheets appeared 2.75 million years ago, when the Northern Hemisphere crossed a new threshold into an ice age, or, more accurately, the start of a long sequence of ice-age cycles. At that time, icebergs began to dump mineral grains and rock fragments from the continents into Atlantic sediments as far south as Newfoundland and France. At the same time, the oxygen-isotope record shows the start of the series of oscillations that have continued right through until the present time (fig. 4.2).

From 2.7 until 0.9 million years ago, most of the glacial cycles occurred at regular intervals of 41,000 years, with a few others at intervals of 22,000 years. Milutin Milankovitch would have been pleased to see this two-thirds of Northern Hemisphere glacial history. His radiation calculations indicated that the 41,000-year and 22,000-year cycles are both strong at the latitudes where northern ice sheets existed, and he had predicted that ice sheets would form during times of low summer radiation at those two cycles. And here they were, some 40 or 50 individual glaciations, each separated by times when the northern ice sheets had melted. The implication of this early pattern was that climate had become cold enough for ice sheets to form when summer radiation was weak yet was still warm enough that the ice melted when summer radiation was strong.

Also evident in figure 4.2 is a very slow drift toward more glacial conditions. The oxygen-isotope technique measures not just ice-sheet size but also ocean temperature. In many regions, the temperature trends are thought to track those of the ice: when the ice sheets grow large, they make climate (and the ocean) colder, and conversely. So the slow drift evident in figure 4.2 is partly an indication that ice sheets were getting larger, but also a sign that the ocean was growing colder.

Nearly 0.9 million years ago, the cooling trend reached another threshold, and a new pattern of ice-sheet variations appeared. Since that time, ice sheets have not melted completely after every individual 41,000-year or 22,000-year cycle as they had previously. Instead, some ice has survived during the weaker summer radiation peaks, providing a base-level amount that could be added to during the next radiation minimum. As a result, ice sheets began to persist for as long as 100,000 years, shrinking back somewhat during the smaller radiation peaks, but then growing even larger. This change in ice response is probably explained by continued cooling of Earth's climate: now the Sun was having trouble melting the ice, when much earlier it had not allowed any ice to grow.

Nevertheless, Northern Hemisphere ice sheets still disappeared for relatively brief intervals. Every 100,000 years or so, the solar radiation peaks are strong enough to melt all the ice on North America and Eurasia very rapidly. These larger radiation maxima occur at times of close alignment of the peaks caused by cycles of tilt (41,000 years) and precession (22,000 years). The most recent of these alignments happened between 16,000 and 6,000 years ago, and the great northern ice sheets melted. Only the small ice sheet on Greenland has endured these major melting episodes.

Over the last 900,000 years, up to nine 100,000-year glacial cycles can be counted in figure 4.2. Superimposed on top of these, and more difficult to see, are 41,000-year and 22,000-year cycles like those that had existed earlier. These shorter cycles did not end when the new cycles developed; instead, they were overprinted by them. Overall, the three cycles have a saw-toothed shape, with a slow buildup to maximum size, but then a very fast interval of melting. Milankovitch would have been surprised to see this pattern. His theory did not predict the longer and larger 100,000-year cycle, and the mechanisms that may link it to changes in orbital eccentricity are still being explored.

If we take a step back from the many wiggles in this long ice-age history, the basic message from figure 4.2 is that the Northern Hemisphere (and the planet as a whole) has been gradually drifting toward a more refrigerated state for the last 3 million years. Before 2.7 million years ago, no large ice sheets existed in the North. From then and until 0.9 million years ago, ice sheets appeared in cycles but then melted away completely; they were probably present during less than 50% of that interval. Since 0.9 million years ago, ice sheets have been present more than 90% of the time and more difficult to melt. Our ice-free condition in the Northern Hemisphere today (excluding Greenland) is part of a very short break in a mostly glaciated world. If this long-term cooling trend were to continue into the distant future, northern Canada and Scandinavia could at some point reach a condition more like that in modern-day Antarctica or Greenland, with permanent ice sheets persisting through any and all orbitally driven fluctuations in solar radiation.

These glacial cycles are superimposed on the longer-term cooling trend in much the same way that daily cycles of heating and cooling are superimposed on the seasonal drift from summer to winter in the higher midlatitudes. As temperatures slowly fall during autumn, birdbaths or ponds may freeze at night, but the ice melts in the midday sunshine. By early winter, the freezes may persist through colder days but thaw during occasional warmer spells. In midwinter, the freezes are deeper and feel permanent. By this analogy, the high northern latitudes have gradually reached the chill of early winter and are slowly heading toward the deep freeze of midwinter. Of course, in the short term (the next several centuries), we face the prospect of rapid greenhouse warming and a big thaw, rather than a slow drift into permanent glacial refrigeration.

Given that the warmth of the last several thousand years is relatively unusual in a world where northern ice sheets are present more than 90 percent of the time, we might well ask what a more typical glacial world is like. The full-glacial world was cold, dusty, and windy, especially in the Northern Hemisphere at latitudes north of about 40°, and especially near the ice sheets. The ice sheet in North America reached as far south as 42°N, more than halfway to the equator, and it accounted for half or more of the "extra" ice present on Earth (the amount in excess of that today). Overall, ice sheets covered 25 percent of Earth's total land surface, compared to about 10 percent today. Most of northern Canada has slowly been scraped clean of its ancient cover of soil by the thick masses of ice that again and again carried off sediment and soil. Toward their margins, the melting ice thinned and flowed down valleys created in preglacial times by running water, but sometimes it rearranged the old drainage systems by scraping out new valleys. The ice also used boulders, cobbles, and pebbles trapped in its lower layers as crude chisels to gouge the hard bedrock.

To the south, in the northern plains and the northern Midwest of the United States, the bulldozing ice sheets heaped up ridges of moraine rubble as high as 40 to 50 meters (150 feet). During summer, melt water from the ice margins flowed in steams and rivers to the south and east. The water picked up sand, silt, and clay and carried them south but left behind coarser gravel, cobbles, and boulders. In the cold of winter, the meltwater streams slowed to a trickle or stopped. Cold winter and early spring winds blew across the debris near and south of the ice margins, picked up the fine sand, the silt, and some of the clay, and blew it across the Midwest. Today these glacial deposits are the legendary sandy loams of the farmlands of the Midwest, among the richest soils in the world.

Europe just south of the Scandinavian ice sheet was similar, yet different. The southern margin of the European ice sheet reached only to 52°N, 10° north of the limit in North America, but it covered all of Scandinavia and Scotland as well as the northern parts of Denmark, Germany, France, England, and Ireland. A smaller ice sheet covered the high Alps of Switzerland, France, Austria, and Italy.

Today the climate of Europe is moderated in winter by large amounts of heat released from the North Atlantic Ocean. The northward extensions of the Gulf Stream carry warm, subtropical water to unusually high latitudes, and these warm waters release almost as much heat in winter as the Sun can deliver through the nearly persistent cloud cover. But when the ice sheets were large, this warm ocean current flowed east toward Portugal rather than north toward Scandinavia, and the cold North Atlantic Ocean filled with sea ice in the glacial winters and melting icebergs in summer.

This icy ocean, aided by cold winds blowing south from the ice sheet in Scandinavia, created polarlike conditions that eliminated the rich forests typical of modern-day Europe southward all the way to the Alps. Only tundralike vegetation survived across the region south of the ice: grass, moss, lichen, and herbs adapted to ground that was hard-frozen in winter but thawed to a depth of several feet in summer. The underlying soil was permanently frozen (permafrost) to great depths. Farther south and east, Europe was grassy steppe, with no trees. Much of central Europe 20,000 years ago was like present-day Siberia.

At least one area became wetter during the glacial maximum rather than drier. Those "youthful-looking" beach deposits in the American Southwest that had hinted at lake levels much higher than today (chapter 3) turned out to be the same age as the ice sheet and a direct result of the ice's effect on atmospheric circulation. Today the wettest weather on the Pacific coast of North America occurs where the winter jet stream intercepts the coast between Oregon and British Columbia (or even Alaska), bringing powerful winter storms and lots of snow. But at the last glacial maximum, the North American ice sheet was so large an obstacle to atmospheric flow that the main path of the jet stream, along with its winter storms, was displaced south to the American Southwest. More winter snow, along with cooler temperatures that slowed evaporation in summer, allowed enormous lakes like the one at Salt Lake City to form.

The subtropics and tropics, although far from the north polar ice sheets, were generally cooler and drier. Deserts expanded, and strong winter winds blew thick clouds of dust westward from the Sahara Desert over the Atlantic Ocean and across to the Americas, southeastward from the Arabian Desert into the Indian Ocean, and eastward from Asia into the Pacific Ocean and on to Greenland, where dust particles accumulated in the ice sheet in layers that survive today. And throughout these large climatic changes in the North, our ancestors continued to evolve toward modern human form.

Chapter Five

ORBITAL CHANGES CONTROL MONSOON CYCLES

ONE OF THE BLEAKEST PLACES on Earth today is hyperarid Sudan, south of Egypt. Dry winds blow sheets and dunes of sand across the landscape, and almost nothing lives there. But satellite photos and images from heat-sensing devices show subsurface traces of streams and rivers that once flowed eastward to join the Nile River in its northward course from well-watered source areas in the highlands of East Africa. Once, this desert area was green, with broad grasslands and tree-lined waterways inhabited by crocodiles, hippopotamus, ostriches, and rhinoceros. Their bones are found in dried-up stream sediments now covered by a thin layer of sand.

Indeed, the entire southern fringe of the Sahara was once green, with the grasslands of the modern-day Sahel region much farther north than now, and large lakes dotting the area. Modern-day Lake Chad in southern Libya is the largest lake in the region today, but not long ago it was ten times larger, based on evidence from freshly deposited lake sediments and notches cut by lake waves into surrounding rocks. Nomadic tribes in the North African desert tell stories of a wetter time in their ancestral past.

Until 25 years ago, geologists were still looking for an explanation for this vast greening of the southern Sahara. Most early attempts had linked this wet interval to the presence of ice sheets in the North. The prevailing hypothesis held that the ice sheets had interfered with the normal circulation of the atmosphere and redirected moist winds southward into a region they do not reach today. The midlatitude storm track had supposedly been positioned much farther south, bringing rains to the arid core of the Sahara Desert and to the grasslands farther south. This explanation had seemed to work (and indeed still does) for the existence of lakes in arid basins of the southwestern United States, and it was used to explain similar climatic changes found in North Africa.

In this case, however, a very reasonable explanation turned out to be wrong. By the 1960s and 1970s, with the discovery and application of radiocarbon dating, the earliest reliable dates of the lake sediments in North Africa were coming in, and the lakebeds did not date to the time when the ice sheets had been largest, near 20,000 years ago. Instead, their ages were closer to 10,000 years ago, by which time the ice had almost completely disappeared from the northern continents. The unavoidable conclusion was that large ice sheets had not caused the filling of the lakes.

Some scientists then proposed the exact opposite view, arguing that the lake levels were higher during warm interglacial climates because the disappearance of the ice sheets and their chilling effect on climate had allowed the atmosphere to hold more moisture and thereby fill the lakes with rainwater. But this suggestion was doomed from the outset. It offered no explanation of why most of the North African lakes had dried out in the last 5,000 years. Because no ice sheets have appeared during that interval, northern ice sheets cannot be the reason for the recent drying trend.

By this point, all explanations of African lake levels tied to ice sheets had effectively reached a dead end. The cause of the greening of the southern Sahara Desert had to lie elsewhere. And as it turned out, the explanation was right there in the tropics—the overhead Sun. In 1981 meteorologist John Kutzbach came up with one of those simple explanations that make other scientists kick themselves for not having thought of it first. He based his hypothesis on the modern-day wet summer monsoon, a phenomenon that is the main cause of rainfall over much of North Africa. To most of us, the word "monsoon" invokes images of torrential rains falling in India and Southeast Asia, an area with the strongest monsoons in the world. But North Africa has its own monsoon, and it operates across the broad Sahel region south of the Sahara Desert. The North African monsoon delivers summer rains that allow grass and some trees to grow in the savanna south of the desert. In winter the rains cease and the land dries out, but the grasses are naturally adapted to seasonal rains and a long dry season. North of about 17°N lies the hyperarid Sahara Desert, beyond the reach of both summer monsoon rains from the south and winter storms from the north.

Kutzbach's simple but elegant insight was that the former existence of extensive lakes and grasslands in the Sahel and southern Sahara can be explained by simply strengthening the modern-day summer monsoon pattern and delivering heavier and more widespread rainfall. To explain this strengthening, Kutzbach called on the same physical process that drives the modern-day monsoon: heating of continents by the strong overhead Sun (fig. 5.1). Heating of land surfaces warms the overlying air, and heated air rises (just as in a hot-air balloon) because it expands and becomes less dense. As the heated air rises, it leaves behind at the surface a region of lower pressure caused by the upward loss of air. To replace this deficit, air moves in from nearby regions. If an ocean is nearby, the air that moves in contains water vapor evaporated from the sea surface. The arrival of this moisture-bearing air from the ocean sets the stage for the wet summer monsoon.

As the moist air flows in from the ocean, it is heated and joins in the upward flow above the hot land mass. But as the rising air penetrates tens of kilometers into the atmosphere where lower temperatures prevail, it cools. Cooled air cannot hold much moisture, and the water vapor condenses into tiny droplets that form

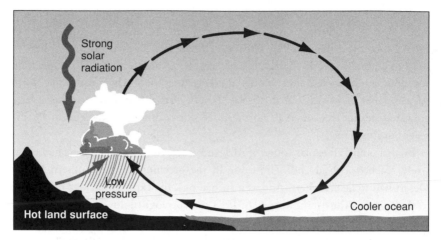

5.1. Tropical monsoons are driven by changes in summer solar radiation at the 22,000-year orbital cycle. Intervals of strong radiation produce heavy monsoon rains that saturate tropical wetlands and produce large emissions of methane (CH_4).

clouds and then raindrops that fall back to earth. You can see this process at work every summer at middle latitudes in the towering cumulonimbus clouds that form high above heated landmasses, usually as scattered clouds that bring afternoon showers. In the tropics, this monsoon circulation is far more widespread and persistent. Many regions receive drenching rains every afternoon and evening because of the midday solar heating.

So the basic operation of the summer monsoon is very simple: strong radiation from the Sun, strong heating of the land, an inrush of ocean air, and monsoon rains. In winter, everything works in exactly the opposite way. The Sun is weak, the land cools off, the overlying air becomes dense and sinks, and the sinking air is dry. This circulation is also called a monsoon—the cold dry winter monsoon. Winter is the dry season in much of the tropics for this very reason.

Kutzbach's hypothesis built directly on this knowledge of modern-day monsoons. Because the stronger radiation of summer (compared to winter) is the cause of the modern-day summer monsoon, he reasoned that past intervals with stronger-than-modern solar radiation in summer should have driven a stronger summer monsoon and filled those North African lakes. And intervals with summer solar radiation values higher than those today have occurred repeatedly in the past.

Solar radiation at tropical latitudes over orbital time scales is controlled by the 22,000-year cycle of orbital precession (chapter 3). At the present time, summer radiation is at a minimum value for that cycle, which implies that our modern-day summer monsoons must also be close to their minimum long-term strength. Kutzbach noted that solar radiation levels were considerably higher than those today at regular intervals of 22,000 years, with the last such time occurring just 10,000 years ago. According to his hypothesis, lakes in North Africa should have reached their largest size near this time, in agreement with the large amount of evidence amassed from radiocarbon dating.

Kutzbach tried out an initial test of his hypothesis using a numerical model of Earth's atmospheric circulation. These models are basically the same kind used in weather forecasting: they incorporate physical principles known to control the circulation of the atmosphere. Of course, making a comparison to weather forecasting always leads to the same wise-guy challenge: "How can you scientists expect to 'forecast' the weather thousands of years ago when you can't even get the weather for this coming weekend half-right?" The response to this (quite reasonable) wise-guy prod is that climate scientists like Kutzbach don't use these models to predict day-to-day weather but instead explore the longer-term average state of the climate system.

The annual climatic cycle provides a useful analogy. No scientist would attempt to use models to make a weather forecast for a specific July day months in the future, yet these same models can reliably predict the *average* conditions during a typical July day. The reason they can do so is that an average day in July is hot for a simple physical reason: the Sun's rays are direct and the days are long. The models do a fine job of simulating the average heating that occurs in response to the strength of the midsummer Sun. If they didn't, they would be totally useless. The same models can also simulate in a general way much of the day-to-day and week-to-week weather variability that occurs during a typical July, but again they cannot make a specific forecast for a particular day.

For the same reason, these models can be used for meaningful tests of the monsoon hypothesis. Kutzbach knew from calculations based on Earth's orbit around the Sun that solar radiation in the northern tropics 10,000 years ago was 8% higher than it is today, and he entered that value as an "initial condition" in running the model to see what kinds of climatic changes it would have caused. In effect, he wanted to find out what an "average July day" would have been like 10,000 years ago (rather than a specific weather forecast for a particular day, which would be impossible). The result from this first model experiment confirmed his hypothesis—stronger solar radiation drove a stronger monsoon that delivered more rain across the North African Sahel in the model simulation.

Kutzbach then turned to geographer Alayne Street-Perrot, who had begun a compilation of radiocarbon-dated lake levels across North Africa. Together they compared the model simulations of precipitation for several time intervals through the last 20,000 years against the radiocarbon-dated evidence from the lakes. The match was excellent. The lakes were low 20,000 years ago during a time when the model simulated a weak monsoon because solar radiation in summer was lower. The lakes were highest nearly 10,000 years ago when stronger radiation drove powerful monsoons. And the lake levels slowly dropped during the last 10,000 years as summer radiation levels waned. Beyond any doubt, the lake levels were following the tempo of the overhead summer Sun, not that of the distant ice sheets.

In the last two decades, the predictions of Kutzbach's hypothesis have been borne out so many times in so many regions that it has come to merit the higher stature of a theory. The basic tempo of dry today, wet 10,000 years ago, and dry again 20,000 years ago holds not just in North Africa but also across a great arc from southern Arabia, India, and Southeast Asia to southern China. Both the field observations and the model simulation shown in figure 5.2 agree that this entire region was wetter 10,000 years ago than it is today because the summer monsoon was stronger than it is now in response to greater amounts of summer solar radiation.

In my opinion, John Kutzbach's theory of the orbital control of monsoons ranks in importance right alongside Milankovitch's theory of the orbital control of ice sheets. Half of Earth's surface lies between 30°N and 30°S, and monsoon changes are the primary orbital-scale control of climate across this vast region. Most of the world's population lives at these latitudes today and has done so since humans evolved. We have lived with these monsoon fluctuations for several million years.

Perhaps the most impressive confirmation of Kutzbach's hypothesis has come from a remote source—ancient air bubbles trapped in the Antarctic ice sheet. Working at station Vostok at the very top of the ice sheet for years in the "warmth" of Antarctic summers (−20°F!), Russian engineers gradually drilled a 3,300-meter sequence of cores down through the entire mass of ice, except a bottom layer left untouched because of concern that an underlying lake might be contaminated by the chemicals used for drilling. Water exists in this improbable location because the pressure of the overlying ice causes melting of the deepest layers of the ice sheet.

Air bubbles trapped in the long column of ice retrieved at Vostok hold a history of changes in the amount of methane in the atmosphere (fig. 5.3). Scientists developed the initial time scale for this ice record based on models that estimate how snow is first compressed and crystallized into ice and then flows slowly down into the ice interior and out toward the ice margins. The methane record produced by this initial time scale showed a strong resemblance to changes in solar radiation at the 22,000-year precession cycle. Based on this evidence, scientists

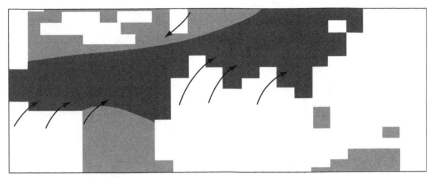

A Model simulation

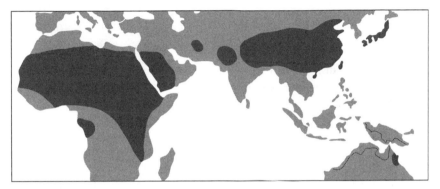

B Observations

■ Greater moisture 9000 years ago than today

5.2. Summer insolation values 8 percent higher than today produced summer monsoons stronger than today nearly 10,000 years ago across North Africa and southern Eurasia. This connection is supported by model simulations (A) and by evidence from Earth's surface (B).

concluded that a major source of the methane changes measured in Antarctic ice must be changes in the strength of the tropical monsoons.

The reason for the connection between monsoons and methane is shown in figures 5.1 and 5.2. When strong monsoon rains fall on tropical wetlands, they flood them with water during the summer season when plants grow. As plants die, they decompose in stagnant water lacking oxygen. Bacteria attack the decaying plant matter and convert its carbon to several products, including methane gas (CH_4). The methane bubbles off from the wetlands into the atmosphere where it stays for an average of 10 years before being oxidized to other gases. This physical connection builds directly on John Kutzbach's orbital monsoon theory: stronger summer radiation causes a stronger monsoon circulation that produces

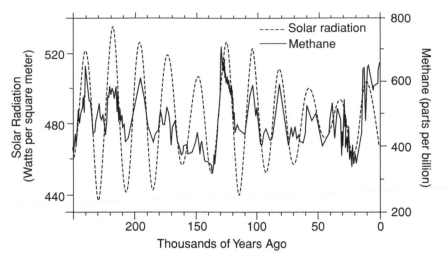

5.3. Releases of methane (CH$_4$) from tropical Eurasia confirm the orbital monsoon theory that stronger solar radiation at the 22,000-year cycle produces stronger monsoon rains, which emit more methane.

more flooding of natural wetlands that emit more methane. It all comes down to a simple mantra: more Sun, more monsoon, more methane.

A second source of methane exists in near-Arctic Siberia and Canada. Natural bogs in these regions also contain plant matter that decays and emits methane gas, roughly half as much of the global total as the amount from the tropical wetlands. Although extensive in area, the Arctic wetlands are frozen much of the year and warm up only enough to allow decomposition of organic matter for a brief interval in summer. Just as in the tropics, however, the key control on methane emissions at the 22,000-year cycle is the strength of the summer Sun, and its effect in warming the land and the wetlands.

The methane trend shown in figure 5.3 was tuned by making small but justifiable adjustments to the original ice-core time scale, tweaking the estimated ages of the methane peaks and valleys by a few thousand years to bring them into alignment with the solar radiation trends. The correlation between the methane record measured in the Vostok ice core and the summer radiation signal calculated by astronomers is obvious. Peaks (and valleys) in both signals occur every 22,000 years, the tempo of the orbital precession cycle. And, to a considerable extent, the amplitudes of the methane peaks track the size of those in the solar radiation signal.

At times, the methane trend was interrupted for 1,000 years or more by abrupt drops in concentration. These decreases occurred at times when the Northern

Hemisphere abruptly cooled and became drier, and the cause of these pulses is not yet well understood. But in between these brief interruptions, the methane trend always returned to the same underlying long-term pattern (more Sun, more monsoon, more methane) that has prevailed for hundreds of thousands of years. Kutzbach's theory is confirmed yet again.

Several scientists have speculated about whether the changes in the monsoon, or in the ice sheets far to the north, could have affected the evolution of humans in Africa. Monsoons affect the size of water sources on which human and other animal life depends. Much of the grasslands and woodlands of tropical Africa lie in the region of monsoon control of summer rainfall.

The distant ice sheets also transmit their climatic impacts to Africa. When the ice sheets are large, winds in North Africa grow stronger, especially in late winter and spring. Dust is always present somewhere in the arid Sahara, and the stronger winds during glacial intervals lift and blow more of it across the desert and out to sea. Regions like the northern limit of the tropical rain forest also become drier in glacial winters.

Some scientists have invoked these climatic changes linked to Earth's orbit as an environmental "challenge" that forced the hominid line to acquire larger brains and a wider range of skills in order to cope and survive. Different variations on this basic idea emphasize different aspects of climate change: the general pressure of a constantly changing environment, the pressure of surviving through the dry parts of the climatic cycles, and so on.

As a nonexpert in human evolution, I have my doubts. I can certainly see how challenges could be a factor in promoting human evolution, but it seems to me that many other, far more immediate and pressing, challenges would have existed. Basic day-to-day life was hard enough: lions crouching in the high grass ready to attack, rhinos making access to the water hole difficult, other humans constantly competing for the same resources, and so on. I find it difficult to see how gradual orbital-scale changes in climate would been a significant factor in human development.

The more abrupt changes that punctuated the slower orbital-scale rhythms are a more promising source of such a challenge. Many of these dry, cool intervals developed in a few decades or even years and then lasted for centuries. Changes this abrupt and intense could have altered in a measurable way how humans lived. But following this line of reasoning, what about the role of good old "weather," the random droughts, floods, heat waves, or chills that even now command so much of our attention for weeks, months, and years? Wouldn't they, cumulatively, constitute a much greater challenge to humans than a slower drift toward wetter or drier conditions? I suspect it will be difficult, if not impossible, to demonstrate convincingly that climate affected human evolution.

Some hypotheses invoke a different impact of climatic change on human evolution—the effect of a gradual drying in fragmenting the forests and forcing (or drawing) hominids into open woodlands and tree savannas where their intelligence and skills developed more rapidly. Here, too, I have my doubts. For one thing, forests survived even in the driest times. I don't see why clever and naturally mobile hominids would not have simply followed the environment they preferred, moving southward as forests shrank toward the moist tropics or to higher elevations as forests retreated up the sides of mountain.

In my (again nonexpert) opinion, it remains an open question whether or not climate change was important in human evolution. It seems to me that once the hominids had come into possession of a larger brain to go along with their already-talented hands and arms, almost anything and everything would have conspired toward behaviors that made use of these enormous survival advantages. Some species of hominids would have made better use of these talents than others and so would have outreproduced and outsurvived the competition, but the possibility of rapid evolutionary change was right there, inherent in those large brains and clever hands. If this is so, it seems to me the biggest mystery lies farther back in time, before those half-human creatures left those haunting footprints in the rain-moistened volcanic ash. When and why did their brains, or those of their ancestors, become large enough to become a partner to those dexterous hands?

Meanwhile, right up until the most recent glacial cycles, our hominid ancestors were still living primitive Stone Age lives and still having no detectable impact on climate. But major changes were imminent.

Chapter Six

STIRRINGS OF CHANGE

IN ALMOST 2.5 MILLION YEARS, hominids had moved only slightly beyond the most primitive level of Stone Age life, adding control of fire and gradually more sophisticated stone tools to their meager repertoire of skills (table 6.1). But once our species appeared, 150,000 to 100,000 years ago, the pace of change quickened, at least by comparison to the nearly undetectable tempo of previous times. By 50,000 years ago, we see the first real evidence of human creative potential emerging in the artistic and aesthetic expression of people who were like us both in physical respects and mental capabilities. They painted amazingly lifelike portrayals of animals on the walls of caves and rock shelters, and they made small statues of human and animal figures. They created jewelry by stringing together shells. They buried their dead with food and possessions for use in a future life. In these changes, we can recognize the early origins of a human "culture" similar to ours today, and we can begin to relate to these people as "us."

Also appearing for the first time are much more sophisticated stone tools designed for specific functions, made using standardized (easily repeatable) methods. And for the first time people began to use bone, a substance much more workable than stone, yet hard enough for many uses. The first needles, awls (hole-punchers), and engraving tools appeared. Needles made possible sewn clothing that fit closely, rather than loosely draped animal hides. With greater protection from the elements, people pushed north into higher and colder near-Arctic latitudes. There they built dome-shaped houses using large mastodon bones for the superstructure and draping the roof with hides of animals for protection from rain and snow.

People also learned how to make rope from naturally available fibers and used it to make snares, lines, and nets to catch small animals, birds, and fish. The hunting-gathering life became a hunting-gathering-fishing life. Bone was shaped into spear throwers to hold stone spear points, and rope was used to help bind spear points to shafts of bone or wood. This new technology produced a lethal and revolutionary new way of hunting, combining a weapon that could kill efficiently with a hand that could grasp it and an arm with the natural range of motion to throw it. Hunters could now bring down larger game from a safer distance, and even huge mammoths could be killed, as cave paintings show.

Inevitably, improvements in technology were combined with improved hunting strategies. People now anticipated the movement of game, used fire and other

TABLE 6.1
Major Stages in Human Evolution

Time of Appearance	Relationship to Us	Mode of Existence	New Tools and Implements	Culture
2,500,000,000 years ago	Our genus *Homo*	Hunting, gathering	Stone Age Primitive spear points Grinding and crushing food Digging roots, tubers	Unknown
500,000 years ago	Our species *Homo sapiens*	Hunting, gathering	Stone Age Less primitive spear points Fire (food and security)	Unknown
150,000 years ago	Fully modern humans ("us")	Hunting, gathering, fishing	Stone Age Advanced spear points Bone needles and awls, rope snares, nets, lines Houses and clothing Primitive metal working	Ceremonial burials Care for sick, elderly Cave paintings Statues, figurines Jewelry
12,000 years ago		Farming	[see chapter 7]	

planning to startle and drive the animals toward intended areas, and communicated with each other as the hunt proceeded. In time the impact of these skills came to be felt in ways entirely different from anything in the prior history of humans or their more distant predecessors. On several continents where humans had not previously been present, major extinctions of large mammals occurred at or very near the time humans first arrived. In Australia people first arrived nearly 50,000 years ago, having by then learned how to make boats that could carry them over the deep channels that still separated Australia from the Southeast Asian mainland during a time of lower glacial sea level. Coincident with the first arrival of humans (insofar as dating methods can tell), a large fraction of Australia's indigenous marsupial population disappeared forever, including a marsupial lion, three kinds of wombat, nine genera of kangaroo, and a range of nonmarsupials including a giant tortoise the size of a Volkswagon "bug" and several kinds of lizards and flightless birds. Every vertebrate species larger than humans disappeared. Nothing even close to this concentrated a pulse of extinction had occurred for millions of years on Australia, despite an ongoing drift of climate toward much drier conditions.

Even stronger evidence ties humans to more recent mass extinctions of large mammals in the Americas. In North America nearly 12,500 years ago, and within the range of accurate radiocarbon dating, many large mammals disappeared in a very short time, most of them within just a thousand years. In all, 33 genera of large mammals died out, leaving only 12 still extant. The list of those lost includes the wooly mammoth, the mastodon, two species of buffalo, four genera of giant ground sloths, one kind of bear, the saber-tooth tiger, the camel, the cheetah, two types of llama, the yak, a giant beaver, several kinds of horse and donkey, the tapir, the peccary, one type each of moose, antelope, and deer, and three kinds of ox. Most of these mammals weighed 100 pounds or more, and most were larger than any surviving counterparts today. One of the ground sloths was up to 20 feet tall and weighed several tons. The Imperial mammoths had curving tusks 12 feet long.

Some of these genera went totally extinct in a global sense, while others survived in parts of the Old World. When Europeans later reintroduced the horse and donkey to North America, they were bringing those creatures back home to the continent where they had first evolved. For tens of millions of years, the grassy plains of North American had been richer in large-mammal game than the modern-day plains of Africa, but now this great collection of animals was abruptly reduced to slightly over a quarter of its original variety.

Some scientists argue that climate change was the culprit. They point to environmental stresses 12,500 years ago tied to changes in Earth's orbit: strong solar radiation that caused hot summers and rapid melting of the great North American ice sheets, weak solar radiation that produced cold winters, and swift northward migration of forest and grassland vegetation into regions from which the ice had melted. One argument was that the grassland habitats favored by many grazing and browsing animals were squeezed out by these coincident and large-scale climatic changes, and that many species became extinct for lack of suitable habitat.

In the 1960s geoscientist Paul Martin first argued the case against a strictly climatic control of the mammal extinctions. He pointed out that more or less the same set of environmental variables had recurred at least four times in the last million years, and yet no mass extinction of mammals had occurred earlier. He concluded that the mass extinctions 12,000 years ago must have been the result of a new factor that had not been present during any of the earlier intervals: predation by humans.

I am certainly no expert on either mammals or cultural anthropology, but I am an expert on climate cycles, and I find Martin's criticism of the climatic explanation absolutely convincing. In fact, it is even more convincing now than when he first presented it. When Martin first argued that humans were the primary cause of the mass extinctions, climate scientists still thought that only four glaciations had occurred. But over the last two decades it has become clear that 8 or 9 glacial cycles occurred in the last 900,000 years, and some 40 or 50 cycles in the full 2.75 million years of the Northern Hemisphere ice ages (chapter 4).

With all of these cycles of ice-sheet growth and melting, the same basic configuration of climatic factors that occurred 12,500 years ago had already happened dozens of times in the past: strong summer radiation, weak winter radiation, rapidly melting ice sheets, and swiftly migrating vegetation. In fact, the extremes of summer and winter radiation had often been larger in earlier glacial cycles than they were 12,500 years ago, with even faster rates of ice-sheet melting and geographic shifts in vegetation, yet no concentrated interval of extinctions had occurred during any of those earlier cycles. Indeed, the number of extinctions near 12,500 years ago in North America rivaled or exceeded the total from *all* of the previous ice-age cycles over 2.75 million years. Why would all of these different types of animals suddenly have become vulnerable to a combination of factors that had not affected them for several million years? To me, the answer is clear. Climate cannot be the primary reason for this unique and highly concentrated pulse of extinctions.

The only possible conclusion left seems to be the one Martin proposed: some entirely different factor must have been at work during the most recent deglaciation. And the most obvious "new" factor was the presence of humans in the Americas. Some scientists believe that humans first arrived almost simultaneously with the pulse of mammal extinctions, while others think they may have come somewhat earlier. Either way, the link in time remains highly suggestive. Part of the answer may also involve an innovation in human technology that appears at the time of the extinctions. A spear point called a "Clovis point" (for the site in Clovis, New Mexico, where it was recovered) was found embedded in the ribs of now-extinct mammals, including mammoths. This evidence tells us that humans were present and were armed with new weapons when the extinctions occurred.

North America is only one of several examples of sudden extinctions that have occurred when humans arrived. An even worse fate befell the mammals of South America, where 80% of the genera (46 out of a total of 58) have disappeared since 15,000 years ago, most during the early part of the interval of human occupation. Again, no such extinction pulse had occurred during the previous millions of years. The same story has been repeated on a smaller scale on islands like Madagascar, Hawaii, and New Zealand. Wherever humans have previously been absent and then appeared, massive extinctions have followed.

In Africa and Eurasia, where humans and animals had co-evolved over very long spans of time, no comparable pulse of extinctions occurred. This observation makes sense because the game hunted by humans had sufficient time in which to evolve strategies for surviving the pressures of human hunting, such as solitary habits and erratic unpredictable migration patterns.

The conclusion that humans caused these massive extinctions in the Americas and on Australia met (and still meets) with strong resistance from many respected scientists. The critics argue that the small number of people present in North and South America 12,500 years ago could not possibly have driven all of those animals to extinction, even using spears armed with Clovis points.

But those early Americans did not have to kill every last member of a species one-by-one with spears to cause the extinctions. Humans hunted in organized groups, using verbal communication and fire to drive animal herds into canyons or arroyos or other topographically constrained features. Some of these drives led to cliffs over which the herds were driven by yells and fire. Some ended in dead-end traps where the confused creatures could easily be killed. Skeletal remains of animal herds dating to this interval have been found piled at the bottom of cliff edges, and often only the topmost skeletons show evidence of butchering for food. The hunting strategies worked so effectively that too many animals were killed to make use of them all.

Recent work by population ecologists shows that large-mammal species can be brought to the point of extinction surprisingly quickly by culling just a few "extra" percent of their population per year. Because most large mammals reproduce slowly, with long gestation periods and few young, they are vulnerable to very small increases in normal rates of mortality. Models that simulate population changes by calculating the long-term effect of rates of birth and death show that increases of just a few percent in normal mortality can cause extinction within a few centuries. Once a species is reduced below the level (density) needed for survival, it is doomed to extinction, even without heavy predation. If these early hunters culled just a few percent more individuals than were normally taken by disease, weather, and old age, extinction would have been inevitable.

Critics also raise the reasonable objection that the fossil record does not allow us to say that all of the mammal extinctions in the Americas happened at exactly the time humans first appeared. It does not. But think back to those volunteers sent looking through sediments for scarce dinosaur remains dating near the time of the asteroid impact (chapter 2). The closer they looked, the tighter (closer in time) the connection with the abrupt meteorite impact became.

Both the remains that mark the first arrival of humans in the Americas and those that record the extinctions of the mammal species are relatively scarce, but the general coincidence of these two phenomena in time is remarkable. I will place my bet that the more scientists look, the closer in time the mammal extinctions will lie to the appearance either of humans or of new hunting technologies. In summary, I reject climatic changes as a primary factor in these extinctions, and I find good reason to think that the cause is tied to humans.

People may have a partly unconscious reason for rejecting human predation as the explanation for the extinctions. Two centuries ago Jean-Jacques Rousseau introduced the concept of the "noble savage": a native people who once lived in total harmony with the environment, killing and eating only what they needed for survival, and nothing more. In a graceful phrase often used in recent years, these people were seen as having "lived lightly on the land." Because Rousseau never traveled to see first-hand the primitive cultures described to him by others, his views were second-hand, simplistic, and romanticized, but they remain influential today. It is unpleasant or even unimaginable to think of the pretechnological peoples of the Americas, Australia, Madagascar, New Zealand, or Hawaii as capable of killing on a large enough scale to cause several of the major extinction episodes in Earth's history. That large an assault on nature is viewed as uniquely a sin of modern society employing modern technology. But the facts say otherwise.

I don't put those early Americans in the company of a man like "Buffalo Bill" Cody, who personally slaughtered tens of thousands of bison from horseback and paraded his achievements in Wild West shows. Nor were they at the level of the passengers who shot buffalo for sport from the windows of moving trains. Like all of their human predecessors, the earliest Americans killed animals primarily for food, for clothing, and for bone tools. The massive extinctions were mainly the unintended consequence of new hunting methods that worked all too well. Nevertheless, the loss of much of this astoundingly diverse range of mammals and marsupials ranks as a tragic chapter in human history. Long before organized human civilization first appeared on the planet, we had greatly impoverished the natural fauna, leaving behind small remnants of a once-rich array of life.

All of this evidence from the last 50,000 years—the cave art, the jewelry, the burial rituals, and even the mammal extinctions—reveals new sets of skills, new kinds of imagination, and new abilities to express and communicate thoughts, feelings, and plans. And yet more than 99.5 percent of the way from the appearance of "man" to the modern era, everyone still lived a nomadic hunting-gathering-fishing life, and everyone still made tools out of stone or bone. Progress toward the modern era had still been incredibly modest, even allowing for the more recent emergence of a range of new capabilities.

But then, about 12,000 years ago, human ingenuity in Eurasia reached a turning point and began to produce an advance so momentous that it put humanity on a gradually accelerating path toward the modern era. This turning point, the discovery of agriculture, took place in the "fertile crescent" of the eastern Mediterranean, an arc-shaped region stretching from modern-day Turkey in the north down through Iraq and into Syria and Jordan in the south. At about the same time, similar changes began in the Yellow River Valley region of northern China.

Humans Begin to Take Control

HUMANS BEGIN TO TAKE CONTROL

By COMPARISON WITH the great sprawl of human and prehuman history across millions of years, the last 20,000 or 10,000 years are just the blink of an eye. The ice sheets were at maximum size 20,000 years ago, a time span only 4 times longer than the interval since Egyptians built the great pyramids. The final remnants of the once-huge North American ice sheet melted nearly 6,000 years ago, just a few centuries before Sumerian civilization developed in modern-day Iraq. Meanwhile, nearly 11,500 years ago, with the shrinking North American ice sheet still covering more than half of Canada, people in the region of southwestern Eurasia called Mesopotamia (modern-day Turkey, Iraq, and Jordan) were slowly taking small steps that would completely transform all of prior human existence. Agriculture was about to develop in that region, and soon after in the Yellow River Valley of northern China.

The agricultural life, often referred to as "pastoral" or "rural," seems a natural world to modern-day city dwellers and suburbanites. A weekend drive in the country is a time for relaxing and "going back to nature." Yet farming is not nature, but rather the largest alteration of Earth's surface from its natural state that humans have yet achieved. Cities and factories and even suburban mega-malls are still trivial dots on maps compared to the extent of farmland devoted to pastures and crops (more than a third of Earth's land surface). So when those people in Mesopotamia 11,500 years ago created agriculture, they set humanity on a path that would transform nature (chapter 7).

This section brings together two of the major themes of the detective story that is central to this book: the failure of natural factors to explain changes in the concentrations of greenhouse gases (methane and carbon dioxide) during recent millennia, and the alternative explanations provided by slowly expanding human activities. Because the methane concentration in the atmosphere is controlled by well-understood processes driven by orbital cycles, its long-term natural behavior can be predicted with the greatest confidence, and departures from the expected behavior caused by humans can be easily detected (chapter 8). The CO_2 system is more complex, but its long-term concentration in the atmosphere is also controlled by orbital cycles, and atmospheric CO_2 values again depart from their natural trend because of human influences (chapter 9).

As human emissions of these gases have steadily increased over the last few thousand years, so too has their combined impact on climate. Had nature remained in full control, Earth's climate would naturally have grown substantially cooler. Instead, greenhouse gases produced by humans caused a warming effect that counteracted most of the natural cooling. Humans had come to rival nature as a force in the climate system. Several lines of evidence in chapter 10 suggest that our releases of CO_2 and methane during the last several thousand years may have stopped a small-scale glaciation that would have naturally developed in far northeastern Canada.

As with any new scientific hypothesis, my colleagues are evaluating this proposal of an early human influence on climate. Two substantive challenges to the original hypothesis have recently appeared in the published literature, both of which have helped me refine and strengthen the original concept (chapter 11).

Chapter Seven

EARLY AGRICULTURE AND CIVILIZATION

AGRICULTURE ORIGINATED INDEPENDENTLY in several regions within the last 12,000 years. The two earliest developments, in the Fertile Crescent region of Mesopotamia at the eastern end of the Mediterranean and in the Yellow River Valley in northern China (fig. 7.1), were to have the largest impact on early civilization. Agricultural discoveries began thousands of years later in other regions, including the Central American lowlands, the high terrain around the Peruvian Andes, and the tropics of Africa and New Guinea. The evidence that agriculture developed independently in these areas rests in part on their geographic separation but mainly on the fact that these areas all have different natural food sources, each of which required a distinct form of domestication.

Studies of remote and isolated tribes who have maintained Stone Age cultures into the modern era have shown them to be world-class botanists within their own habitat. Most can distinguish hundreds to thousands of plants, an expertise rare among modern-day specialists in botany. Their motivation for acquiring this knowledge is obvious: they needed to gather food for survival, and they could not afford to make mistakes. Among edible items, it pays to know which are nutritious and good tasting, and which are not. It also pays to know which roots or nuts or berries or mushrooms are poisonous and which are not. I know a colleague who almost died when he made a wrong choice while foraging for mushrooms. These food choices made long ago were not trivial, and the Stone Age people of the Fertile Crescent region were certainly botanical experts.

From at least two perspectives, the switch from the hunting-fishing-gathering life to agriculture was not inevitable. For one thing, hunter-gatherers derive food from many sources, and the wide variety of easily available plants and animals in regions like the Fertile Crescent naturally promoted nutritional balance. By comparison, overreliance on food sources from just one or two crops can cause malnutrition because of loss of sufficient protein or fat. From this perspective, it did not necessarily make sense for people to rely increasingly on just a few crops after 12,000 years ago. Studies have also shown that less energy is expended harvesting some wild grains in the Near East today than is cumulatively used in planting, tending, and harvesting their domesticated equivalents. We can assume that these early people were constantly making sensible, prioritized choices of how to get the most food in the least time and for the smallest expenditure of energy.

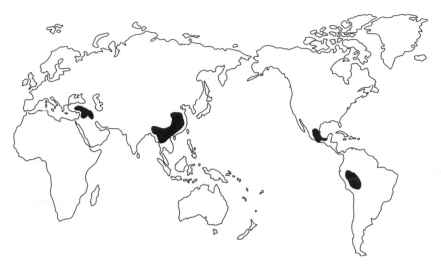

7.1. Agriculture originated independently in several areas shown in black, with the earliest known occurrences in the Fertile Crescent region of the Near East and in northern China.

Even though the origin of agriculture was not inevitable, several factors caused people living in the Fertile Crescent to begin altering their daily routines in ways that would gradually develop into the practice of agriculture. Growing wild in the semi-arid grasslands of this region was an unusual variety of edible wild grains that were ideal for domestication (table 7.1). These included several kinds of cereals: two kinds of wheat (emmer and einkorn), barley, and rye, all of which are easily gathered sources of carbohydrate. Also available were peas and lentils (beans), both good sources of protein. This range of natural bounty was unique to this one region.

Part of this bounty was a result of the semi-arid climate of the Near East. In regions with long dry seasons, where most of the annual vegetation dies each year, plants use their energy to produce seeds for reproduction, and those edible seeds are convenient food sources for humans. By comparison, the much larger amount of vegetation stored in well-watered forests provides little that is edible, and deserts are nearly barren. As a result, people in the Fertile Crescent could derive a portion of their food needs from crops but still augment their nutrition intake by gathering wild seeds, hunting, and in places fishing. With this combination of resources, they could avoid malnutrition. Because of this variety and natural abundance of wild foods, several technical innovations appeared early in this region (or were quickly adapted from elsewhere) by the time of the start of agriculture: primitive stone sickles with which to cut grain, woven baskets in which to carry it, and mortars and pestles for grinding it.

TABLE 7.1
First Appearance of Domesticated Crops and Animals, and Tool Technologies

Years Ago	Near East	East-Central Asia	Americas	Tools (Eurasia)
12,000	Dog		Dog	
11,000	Goat, sheep Wheat, barley Pea, lentil Pig			
10,000	Rye Cattle	Pigs Chickens	Bottle gourd Squash Chili pepper	STONE
9,000	Flax	Millet	Avocado Beans	and
8,000		Bottle gourd Dog Water chestnut	Corn Llama, alpaca	BONE
7,000	Date palm Vine	Mulberry Rice Water buffalo		
6,000	Olive Donkey	Horse Cattle	Cotton	————
5,000	Melon, leek Walnut	Onion Camel	Peanut Sweet potato	BRONZE
4,000	Camel	Garlic		————
3,000			Potato Turkey	IRON
2,000				

And so a transition to agriculture gradually began to take hold in this region. At first, people picked wild grains and legumes because they were available, tasted good, and during a particular season were one ingredient among many sources of food. This was "gathering," not agriculture. But by 11,000 years ago, cereals related to wild varieties of grain began to appear well outside their natural distributions in areas like the Tigris-Euphrates floodplain as clear evidence of human intervention and manipulation. Very slowly, these preserved grains become larger than their original wild form, by as much as a factor of ten for crops like peas. People repeatedly picked the largest grains or vegetables because more food could be gathered in less time by doing so. This selection process was initially unconscious, simply a matter of common sense and efficient use of time, but it gradually produced results similar to what agricultural specialists have consciously done in recent centuries to develop improved strains of crops.

Grains that were picked but uneaten would have been scattered around campsites or early farm villages. Some seeds would have passed intact through human digestive systems and ended up in nearby waste piles and garbage dumps. Some grains would have sprouted nearby during the next growing season in favorably moist and fertile locations, handy for use in that year's food supply. Slowly, people began to make these hand-selected wild grains a larger portion of their annual diet, and at some point they started to use sticks to poke holes in the ground to plant them, even though still heavily dependent on hunting and gathering for most of their food. The grains were easy to plant and grew to maturity quickly, usually within three months.

Nearly 10,500 years ago, the first evidence of permanent settlements appeared, farm villages with populations in the hundreds. The earliest remains are mud-brick houses, surrounded by animal bones that indicate year-round butchering at the same location. These dwellings are direct evidence that people had largely made the conversion to farming, staying in one place to tend their fields and to watch over their food stores in winter. Slowly, over the centuries, humans and the grains they grew gradually became more interdependent. As people continually selected the largest seeds, their crops evolved into hybrids incapable of surviving in the wild against natural competition from other vegetation. Freed from the need to move constantly and carry their young, and with dependable food sources available, people could have children more frequently, and populations began to rise.

Another major advantage of the Fertile Crescent region was that it was the natural home to many types of wild animals that proved easy to domesticate. As Jared Diamond noted in *Guns, Germs, and Steel*, many more types of mammals exist than can easily be domesticated. Some are by nature too wild or skittish or solitary, and some are simply too small to be of much use. The Fertile Crescent was again

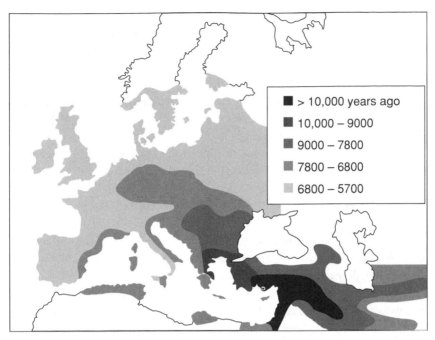

> ■ > 10,000 years ago
> ■ 10,000 – 9000
> ■ 9000 – 7800
> ■ 7800 – 6800
> ■ 6800 – 5700

7.2. Plants that were first domesticated in the Fertile Crescent region of the Near East more than 10,000 years ago gradually appear in lake sediments across Europe in subsequent millennia.

fortunate in having the wild predecessors of goats, sheep, pigs, and cattle. The hunter-gatherer technique of flushing and driving game in the wild probably evolved by degrees into herding of semiwild flocks and later to full domestication.

By 9,500 years ago, goats, sheep, pigs, and cattle had all been domesticated (table 7.1), as well as dogs to guard villages, scavenge debris, provide bedtime warmth, and, on occasion, be a source of food. The milk, cheese, and meat from domesticated animals were major sources of protein and fat to supplement carbohydrates and protein from grains. Nutritionally, the Fertile Crescent provided everything necessary for humans to adapt a new kind of life based entirely on agriculture.

Gradually, knowledge of agriculture spread from the eastern Mediterranean (fig. 7.2). The earliest presence of agriculture in any region is marked by the first occurrence of domesticated cereal and legume grains in radiocarbon-dated lake sediments from regions where these plants did not grow in the wild. By 9,000 to 8,000 years ago, this Fertile Crescent package of crops had reached eastward into India and westward into modern-day Greece. By 7,000 years ago, it had spread southward into Egypt and Tunisia, westward into southern Italy and Spain, and northward into Europe as far as Germany. By 6,000 years ago, farming was present in all of Europe where it is practiced today. In a regional sense, the spread of agriculture was slow, but on a site-by-site basis the switch to agriculture was often rapid. With human help, cereals and other crops adapted to semi-arid climates of the Near East grew abundantly in the moisture-rich climate of Europe. Because wild cattle and pigs (boars) were plentiful in central and northern Europe, they became the primary kinds of domesticated livestock, rather than the goats and sheep common in southern Europe.

Southeast Asia, the second major region where agriculture originated, spans a wide range of environments. In the cool, seasonally dry latitudes of northeastern China along the floodplain of the Yellow River and in the higher and more arid lands to the west, people grew two kinds of the grain millet. The earliest remains of these crops date to 9,500 years ago. Also found are the bones of domesticated chickens and pigs, as well as stone tools and pieces of pottery. In the warmer, wetter regions of tropical southern China and westward into Burma and Thailand, rice farming developed in a series of steps, beginning with the gathering of wild strains prior to 8,000 years ago, with intentional planting developing over the next two millennia. Because wild rice occurs only in uplands, the earliest rice farmers probably poked holes in relatively moist ground with sticks and planted grains. By 5,000 years ago, controlled irrigation of fields for rice growing had begun.

Other agricultural regions were making similar advances. Between 9,000 and 8,000 years ago, the distinctive package of Fertile Crescent crops arrived in the Indus River valley of western India, and locally domesticated crops like cucumbers and sesame were added to the mix. Rice-farming techniques from Southeast Asia reached the Ganges Delta region of east-central India by about 3,000 years ago. Other people elsewhere in the world were also independently discovering agriculture. In the Americas, squash was grown almost 10,000 years ago, and avocados, beans, and corn by 9,000 to 7,500 years ago. Potatoes and tomatoes were also domesticated in the Americas.

As the millennia passed, the impact of agriculture on people's lives continued to expand and diversify. Initially people grew fibrous plants like flax, cotton, and hemp for clothing, blankets, and rope. Later they obtained wool from sheep, goats, llama, and alpaca, leather from cattle, and silk from silkworms, and they grew bottle gourds in which to carry water.

Beginning nearly 6,000 years ago, and continuing through Roman times, another important threshold in the impact of agriculture on humanity was reached. Agriculture and other innovations began to spur the growth of major civilizations, which in turn had a growing impact on the environment. One major innovation was in metallurgy. Humans in many regions were finally at the point of leaving behind the Stone Age tools they had used for over 2 million years. This process probably began in part with a simple observation around campfires: some of the rocks used to line the fire hearths contained small veins of shiny metals that melted at relatively low temperatures, ran out in liquid form onto the ground, and solidified when the fire weakened. In some regions, relatively pure veins of metal were readily apparent in rock outcrops. The obvious next step was to melt and refine copper and tin and later the alloy they created, thus initiating the Bronze Age 6,000 to 5,500 years ago in northern China. Further advances led to the start of the Iron Age by 3,300 years ago in southern China.

Another important advance, the domestication of water buffalo and horses nearly 6,000 years ago, provided animals capable of pulling plows with much greater power than humans could possibly manage. With strong animals now harnessed to durable metal plows, the labor involved in farming was transformed, and ever-greater areas could be plowed and brought into production.

Irrigation also became widespread during this interval. Small, primitive irrigation canals in the higher terrain of modern-day Turkey date back to 8,200 years ago, and the technique spread down into the Tigris-Euphrates River valleys between 7,300 and 5,700 years ago. Previously, farming in the generally arid lowlands near these rivers had depended on rivers flooding to just the right degree: enough to provide occasional water to nearby fields, but not so much as to inundate them for long intervals or erode the soil. Obviously, nature was not always so cooperative. Now, greater control of water meant that farmers were free from most of nature's whims. By 4,000 years ago, irrigation had come into wide use in rice-growing regions across Southeast Asia.

Technological and agricultural innovations continued. The wheel, invented nearly 6,400 years ago in the region west of the Black Sea, spread through all of Eurasia by 5,000 years ago. By 5,500 years ago, people had learned how to plant cuttings or seeds and create orchards of olives, figs, dates, and grapes. Later, using the more demanding technique of grafting, they planted orchards of apples, pears, cherries, and plums. Former "weeds" like oats, turnips, radishes, lettuce, beets, and leaks were domesticated and cultivated. By 2,000 years ago, virtually every major food crop we know today was under cultivation somewhere in the world.

As a result of these innovations, food production increased enormously, as did population levels. A global population of a few millions or tens of millions at the start of the Bronze Age 6,000 years ago grew to 200 million by 2,000 years ago, the time of the Iron Age. Innovations in technology and plant cultivation permitted more and more people to be fed, and surpluses were stored away for lean years.

In regions where food production and populations expanded earliest, advanced civilizations developed. The growing populations began to cluster first in villages, then in towns, and later in cities of increasing size. Complex social and political structures developed under centralized ruling power to manage large-scale irrigation projects and distribute food. Social stratification increased, with more marked distinctions between the affluent and the poor. Hereditary rulers used the power of taxation to provide for standing armies. The state also paid metal workers to improve technology and scribes to keep track of commercial transactions. Those who farmed part of the year under the direction of the state became available during the fallow season for projects under state direction, including the earliest construction of monumental architecture. Not by coincidence, the earliest civilizations arose in the same region of the Near East as the initial discovery of agriculture, beginning with the dynastic civilization of Sumeria nearly 5,000 years ago.

Similar innovations and advances continued in China and Southeast Asia: soybean and pea farming; citrus fruit, peach, and apricot orchards; domestication of ducks and geese; tending of mulberry trees as habitat for silkworms; and development of the world's earliest Bronze Age and then Iron Age metallurgy. Soon after 6,000 years ago, local cultures in China began to coalesce into larger aggregates, with fortified towns appearing between 5,000 and 4,000 years ago. China was unified into dynastic states between 4,000 and 3,000 years ago. The appearance of monumental constructions (palaces, defensive walls, and canals) again indicates a surplus of labor from farming people during the fallow season. By 2,200 years ago, all of China was unified under the Zhou and then the Qin dynasties.

In Egypt, a highly stratified society came into existence nearly 5,500 years ago. As the climate became extremely arid and the Nile River became a less reliable source of water because of the weakening monsoon (chapter 5), mechanical methods were developed to lift water to the nearby fields. Food-based wealth allowed construction of the great pyramids. Several millennia later, the engineers of the Roman Empire constructed buildings, aqueducts, and roads, many of which still stand today because of remarkably high-quality cement.

By 4,500 years ago, advanced civilizations with populous cities developed in the floodplain of the Indus River and its tributaries. These people were skilled in making Bronze Age tools and traded products with their neighbors. They lived in homes made of kiln-fired brick along well-planned city streets with well-maintained drains. They developed a pictographic script and were skilled in sculpture and in sciences based on mathematics. Similar advances soon followed in east-central India along the Ganges River.

In all of these regions, writing either developed independently or was adapted from elsewhere, at first primarily to keep track of distributions of water for irrigation and food for human consumption. Written records appeared in Sumeria (modern Iraq) by 5,000 years ago, in the Indus valley by 4,000 years ago, and in

China by 3,300 years ago. Long before free prose or poetry was written down, humans kept extensive records of the amount of food people produced or received, and the tools they were given for farming. Also recorded in varying degree were state functions, births and deaths, and astronomical events.

The world's major religions came into existence during this interval. Some pretechnological societies had focused their spirituality on the animals they hunted, partly as an attempt to acquire the strength of the creatures they pursued, and partly to atone for the act of killing. Other early societies focused their devotion on the Sun and its seasonal motions through the sky. Life in these earlier times depended almost entirely on the Sun, in part for the warmth it provided, but mainly for the growing season it made possible. Farmers welcomed the slow rise of the Sun higher in the sky after the winter solstice, with its promise of a new planting season to come. Early solar observatories have been found on most continents, and priestly castes probably gained prestige from using them to forecast movements of the Sun and planting seasons.

All the major religions of the modern era came into existence in the interval between 3,200 and 1,400 years ago. The Old Testament mainly relates events from the time of Moses nearly 3,200 years ago until the century before Christ. Most of the major religious figures of eastern Asia were born within a few decades: in 604 BC, Lao Tzu, the shadowy figure behind Taoism; in 570 BC, Siddartha Gautama the Buddha, born in Nepal near the northern border of India; and in 551 BC, Confucius, whose practical ethical precepts guided China into the early 1900s. Later, Christ's birth began the era by which the West marks its calendars, and Mohammed's birth followed in AD 570. According to Huston Smith (*The World's Religions*), some historians of religion link the relatively condensed timing of these religious awakenings to the social inequities and injustices produced by agricultural wealth. The founding figures of most religions showed deep ethical and moral concerns for the lives of people in general, but specifically for the plight of the disadvantaged in increasingly wealthy societies.

This panorama of human ingenuity that developed in Eurasia (and elsewhere) after 6,000 years ago is truly astonishing, especially in comparison to those prior millions of years with so little change. But the agricultural innovations, the plants domesticated for crops, the trees and vines planted for orchards, the metal plows and axes, the domestication of oxen and horses, and the control of irrigation were producing other kinds of changes that scientists have only recently begun to examine closely. For the first time in human history, we were becoming a major factor in altering Earth's natural landscapes. Neil Roberts, in *The Holocene*, titled a chapter on the interval subsequent to 5,000 years ago "The Taming of Nature." Land had now become a resource to be exploited, and with this approach came the first serious environmental degradation from human actions. For the first time, human impacts were large in scale: cutting of forests to open the land to crops and pastures, erosion of hill slopes where deforestation and overgrazing destabilized soils, and influxes of eroded river silt and mud that clogged coastal deltas. The environmental impact of humans on the landscape in populous regions was growing.

Changes were also underway in environments far from the wooded regions of southern Eurasia. People living in semi-arid climates began to adopt a different style of life as pastoral nomads, driving herds of domesticated animals in a continuing search for fresh grass. This way of life developed in the Central Asian steppes nearly 4,000 years ago. The vulnerability of these people to drought was later to play a role in the emergence of the Huns, Turks, and Mongols as some of the great invading warriors of history. In Arabia, and in Africa along the southern rim of the Sahara, camels became the means of conveyance for a different kind of nomad.

In the Americas, a relatively early start in agriculture led to gradual but impressive progress. The domestication of corn (maize) was a strikingly successful example of slow plant selection by humans. Teosinte, corn's predecessor plant in the wild, had tiny ears (cobs) about an inch (2–3 cm) long. Slowly, over several thousand years, millions of decisions by individual farmers to select and plant kernels from the larger forms of this plant resulted in corn close to the size we eat today. Cotton, potatoes, peanuts, and several kinds of beans were also domesticated and cultivated in the Americas. The cumulative achievement of early Americans in ongoing plant selection for agriculture rivaled those in Eurasia.

In the Americas, animals suited for domestication and harnessing were rare, some of the more promising candidates having been killed off in the extinctions nearly 12,500 years ago. Llama and alpaca were suitable for wool and occasionally as pack animals, but large draft animals like horses and oxen were not available. Still, after 3,000 years ago, large-scale clearance of forests for agriculture had occurred in the Yucatan Peninsula of Central America, where the monumental structures of the Mayan civilizations were later to rise. In this region, people double-cropped plants grown in raised fields among swampy lowlands. Recent investigations also confirm the presence of surprisingly large farming populations in the Amazon Basin just prior to the arrival of Europeans.

Agriculture also got under way early and independently in Africa and in New Guinea, where tropical rain forests and savannas yielded millet, oil palm, sorghum, and peas. In the western South Pacific, the influence of southern Chinese agriculture and culture spread southward across much of Polynesia in two phases after 3,800 years ago, bringing breadfruit, yams, potatoes, taro, and bananas along with dogs, chickens, and pigs to these islands through a series of ambitious sea voyages.

Technological progress relevant to agriculture (and eventually to industry) continued during the last 2,000 years. In China, coal was mined and used for fuel and cooking during the last 3,000 years. Water wheels were in wide use by Roman times, and the Romans also developed remarkably advanced techniques for making plaster and cement. By medieval times, 1,000 years ago, water mills and tidal mills were used for grinding grain and windmills for drawing water in much of Europe. Well before the industrial era, these sources of natural power drove mechanical devices in blast furnaces and textile production, and were used to saw wood.

Most of the technological flow in the West during the millennium prior to AD 1500 was directed from more advanced Muslim countries to the less advanced European nations, although the Mongol invasions during the 1200s had destroyed much of the irrigation infrastructure in the regions of modern-day Iran and Iraq. The vulnerable semi-arid lands of the Near East also suffered far more environmental degradation than the better-watered and more resistant humid regions. In addition to erosion of uplands and siltation of river mouths, irrigation slowly increased the salt content of many agricultural areas in the Tigris-Euphrates valleys and elsewhere, until they had to be abandoned.

Increasingly, technology was also being put to other purposes. Gunpowder was invented in China well before 2,000 years ago, although that country later turned away completely from its remarkable record of technological innovations after an internal power struggle in the 1400s. The Muslim nations developed their own explosives by AD 1100, and cannon by the 1340s. At least one of the major inventions of the last millennium was peaceful in intent: Gutenberg's printing press invented in 1455. It helped make possible the dissemination of knowledge during the Reformation and Renaissance of the Middle Ages.

Well before the industrial era began in the late 1700s, humans had come a very long way from those first sporadic attempts to plant a few grains of wheat and barley some 12,000 years earlier. Within the last several thousand years, we had become a force capable of transforming the very look of the landscape. And, as the next three chapters will show, we had also become a factor in the operation of the climate system.

Chapter Eight

TAKING CONTROL OF METHANE

SEVERAL YEARS AGO, just as I was about to retire from the Department of Environmental Sciences at the University of Virginia, I noticed something that didn't make sense. The methane concentration in the atmosphere during the last 5,000 years had risen, when everything I knew about the climate system told me that it should have fallen.

My expectation that the methane concentration should have been dropping came directly from John Kutzbach's theory of the orbital control of monsoons (chapter 5). Summertime solar radiation in the northern tropics reached the most recent of its many 22,000-year peaks nearly 11,000 years ago and since then has declined to a minimum value today, exactly half a cycle later. According to Kutzbach's theory, this gradual drop in solar radiation should have driven a corresponding decrease in the intensity of the tropical summer monsoon, which should have caused natural wetlands in the tropics to dry out. As the wetlands dried out, they should have emitted less and less methane as the millennia passed.

At first, the methane concentrations did follow the trend predicted by the mantra suggested by the orbital monsoon theory: more sun, more monsoon, more methane, and conversely. A high-resolution record from an ice core in Greenland shows a methane peak nearly 11,000 years ago, the same time that summer (mid-July) radiation reached a maximum (fig. 8.1). Then, as summer radiation values began to decrease from 11,000 to 5,000 years ago, the methane concentrations also fell, as expected. So far, so good.

But 5,000 years ago, the expected relationship abruptly broke down. Solar radiation continued its long-term decrease, but the methane concentrations began to rise. This anomalous increase then continued for the next 5,000 years, right up to the start of the Industrial Revolution some 200 years ago. By that time, methane values had risen all the way back to the level reached during the preceding monsoon maximum, yet solar radiation had fallen to a long-term minimum in the precession cycle.

This just didn't make sense to me. The methane increase during the last 5,000 years was in direct violation of the "rules" that had held up so well over the preceding hundreds of thousands of years. How could a guiding principle that had operated so effectively for so long have suddenly and totally collapsed?

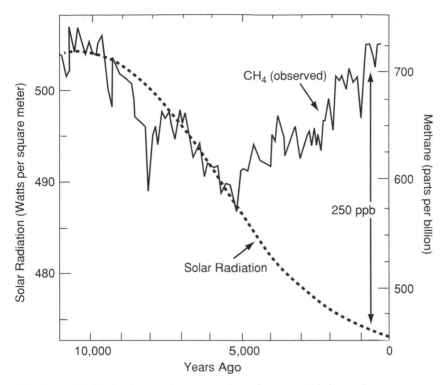

8.1. Natural (orbitally driven) changes in solar radiation caused the methane maximum 11,000 years ago and the decrease until 5,000 years ago, but humans account for the anomalous increase since that time.

One obvious thing to check was the status of the two major methane-generating reservoirs, the tropical and boreal wetlands. Had one or both of them begun to disobey the normal (natural) control by solar heating, even though they had behaved so predictably for so long? But this was not the case. During the last 5,000 years, the tropical wetlands had reacted to the Sun's radiation just as expected. An enormous swath of Southeast Asia, India, Arabia, and North Africa that had been much wetter 10,000 years ago had continued drying out in recent millennia (see fig. 5.2). Individual lakes in the tropics fluctuated briefly between higher and lower levels, but the monsoon region as a whole had followed the Sun's direction and continued to dry out as the millennia passed. In fact, lake levels and other indicators confirm that conditions today in this region are just about as dry as they ever get during the long-term climatic cycles. Because natural wetlands in the tropics have been shrinking just as expected, they cannot explain the increase in methane.

With the tropical wetlands ruled out, the boreal (circum-Arctic) wetlands are left as the other natural methane source capable of explaining the methane trend during the last 5,000 years, and some scientists initially proposed that this region accounted for the methane increase. This idea seemed to make sense because peat bogs have been slowly forming across northern Eurasia and North America for the last 10,000 years, and peat-bog wetlands generate methane at high latitudes. But this explanation was soon eliminated by a clever piece of scientific detective work.

The evidence against a growing Arctic source of methane in the last 5,000 years comes from differences between the amount of methane trapped in air from Antarctic ice versus the amount sealed in Greenland ice. Methane stays in the atmosphere for an average of about 10 years before it is converted to other gases. Because of this relatively quick removal, and because most methane sources are in the Northern Hemisphere (both in the tropics and the Arctic), more methane survives the trip to Greenland than makes it all the way to the South Pole. As a result, concentrations in Greenland ice average about 5–10% higher than those in Antarctic ice.

Over time, the difference in methane concentrations between Greenland and Antarctic ice changes in response to the relative strength of the tropical and Arctic sources. If Arctic wetlands are releasing more methane, then the difference between the ice sheets grows larger, because methane from Siberian wetlands travels relatively quickly to Greenland and much of it is still intact when the air bubbles are trapped in the Greenland ice sheet. In contrast, less methane survives the long trip to Antarctica and is sealed in its ice.

The pattern is different if the methane comes from north tropical sources. Because the north tropics are closer to the equator, methane released from that region travels nearly comparable distances to both ice sheets and along the way oxidizes by nearly equal amounts. Because the main tropical sources are north of the equator, slightly more tropical methane still gets to Greenland than to Antarctica, but the differences are not nearly so large as they are if the methane comes from Siberia. As a result of these different pathways, the methane-concentration difference between Greenland and Antarctica ice grows larger when Siberian sources are increasing relative to tropical sources, and smaller when tropical sources are increasing relative to Siberian sources.

During the last 5,000 years, the interval when the anomalous rise in methane has occurred, the difference in methane concentration between air bubbles in Greenland and Antarctic ice has decreased. This evidence indicates that the amount of methane coming from Siberian wetlands has been stable or decreasing. The methane needed to explain the anomalous overall increase clearly did not come from boreal sources.

This seems like a dead end. Tropical wetlands are ruled out as the methane source because they were drying out, and the decreasing methane gradient between the hemispheres rules out circum-Arctic wetlands as the source. Neither of the two major natural wetland regions was the source of the methane increase. To compound the mystery, the methane-gradient evidence also indicates that the source of the methane increase had to have been the tropics, the region where the natural wetland sources were drying up.

By simple process of elimination, the only possibility left is that the extra methane coming from the tropics must have been generated by a new source that had never before been significant during the previous several hundred thousand years of Earth history. And the new source that fit that description was humans. I first made the case for a human explanation of the methane increase over the last 5,000 years in a paper published late in 2001, with University of Virginia undergraduate student Jonathan Thomson as my co-author.

This was not the first time that scientists had postulated a preindustrial methane signal from human sources. In several papers, scientists had concluded that the rapidly growing human populations in Southeast Asia during the last several centuries prior to the industrial era played a role in the slowly increasing methane concentrations observed in ice cores. But now we were proposing human impacts on quite different scales: beginning fully 5,000 years ago, and growing to a size that rivaled the natural range of variation over long orbital cycles.

But our conclusion, arrived at by process of elimination, was only a start, a basic concept that needed to be backed up by firm evidence. Could humans really have affected methane emissions on such a scale? For humans to be the cause of the anomalous methane increase, some kind of methane-generating activities would have to have begun nearly 5,000 years ago and then gradually increased to much larger size by the time the Industrial Revolution started.

The most likely explanation lay in several innovations that had come about as a result of the origin of agriculture, the most important of which was probably the diversion of river water to irrigate rice (chapter 7). Dry-adapted strains of wild rice had been gathered, grown, and harvested in upland areas of Southeast Asia since before 7,000 years ago. During the same interval, the technique of irrigation had been discovered in the Near East, and the practice had spread eastward to Southeast Asia and southern China. Nearly 5,000 years ago, irrigation was first used to flood low-lying terrain and grow wet-adapted strains of rice in China and Southeast Asia. In effect, irrigation was being used to create artificial wetlands in which rice could be grown. In these wetlands, vegetation (both rice and weeds) grew, died, decayed, and emitted methane, but this methane was from human, rather than natural, sources.

Other human activities also generated methane. Livestock had been domesticated thousands of years earlier in the Near East, and just before 5,000 years ago in Southeast Asia (table 7.1). Livestock generates methane at both ends, both from the manure produced and from gaseous emanations (burps and belches) from stomachs. Animals that feed on raw vegetation have complex digestive systems, with multiple stomach chambers to accomplish the conversion of plant cellulose to digestible form, and microbes in their guts to do the actual work. As herds of domestic livestock grew larger, so did their methane emissions.

Another source of methane was biomass burning. As humans began to clear forests and to burn grasslands to open up space for early agriculture, the burning added more methane to the atmosphere. Still another source of methane was human waste. As human populations grew because of stable food sources, their own waste contributed methane directly to the atmosphere. So humans and their activities—rice irrigation, livestock emissions, and biomass burning—seem to be promising explanations for the anomalous methane increase in the last several thousand years.

But which of these sources caused the changeover from a methane decrease to an increase nearly 5,000 years ago? I think irrigation is the most likely culprit. Most of the other factors—human waste, livestock emissions, and biomass burning—would have been tied fairly closely to the number of humans living, and human populations are thought to have increased relatively gradually over those thousands of years. Yet the departure of the methane trend from its natural path nearly 5,000 years ago was rather abrupt (fig. 8.1). Within the dating uncertainties typical of archeological records, this was the time when irrigation was first used to grow rice in Southeast Asia. I find it easy to imagine that this valuable innovation would have spread rapidly among a population that already numbered in the millions or tens of millions. Large lowland regions in Southeast Asia could have come under irrigation and begun releasing methane.

The continued growth in the size of the methane anomaly shown in figure 8.1 is consistent with the history of irrigation and population growth in Southeast Asia. Irrigation first appeared in southern China and northern Indochina, but by 3,000 years ago it had spread westward to the Ganges River valley in east-central India. Across this vast region, low-lying areas immediately adjacent to rivers and streams would have been the first to have been utilized for irrigation. Later, regions lying farther from the major sources of water would have been brought under cultivation, but only with canal systems that required a considerably greater expense of labor. Later still, during the last 2,000 years, people began to construct hillside rice paddies that further expanded the rice-growing area, but again only with ever-greater amounts of human labor.

This progressive geographic expansion of rice farming is consistent with the continuing rise in the methane trend over 5,000 years. As noted earlier, methane remains in the atmosphere for only 10 years before it oxidizes to other gases. As a result, the rising methane anomaly through the last 5,000 years requires a continually increasing methane input century by century from ever-larger areas of irrigated land. Compared to the first detectable methane anomaly of about 10 parts per billion nearly 5,000 years ago, the anomaly of about 250 parts per billion reached by 500 years ago (fig. 8.1) requires roughly 25 times as large a source of emissions.

During this interval, which spans both immediate prehistorical time (5,000 to 2,000 years ago) and historical time (since 2,000 years ago), human populations are estimated to have doubled roughly once every 1,000 to 1,500 years. If 4 to 5 such doublings occurred over 5,000 years, populations would have increased by a factor of 16 to 32. The 25-fold increase in methane release required by the above estimate falls within the same range as the increase in populations being fed by rice, in agreement with the hypothesis that humans are the most plausible explanation for the anomalous methane trend.

Another way to estimate the size of early methane releases by humans is to work backward from the present day to the time just before the start of the industrial era (about AD 1700). Could the methane releases have grown large enough by 1700 to account for the estimated size of the methane anomaly in figure 8.1? The amount of methane emitted by 1700 can be estimated from a simple calculation based on a proportional relationship. We can assume that the size of methane emissions by humans in 1700 was linked to the number of humans alive then in the same way as today's methane emissions are tied to the number of people living today. In making this calculation, it is obviously necessary to exclude new (industrial-era) forms of methane emissions from human sources (such as burning of natural gas) that did not occur in 1700. We need to focus only on those sources that have existed for centuries.

The first step is to sum up all the methane emitted by activities linked to the 6 billion people living today: rice irrigation, livestock tending, biomass burning, and human waste. For the next step, history gives us a reasonably accurate estimate of the number of humans alive in 1700 (650 million people). Because the 6 billion humans living today emit a known amount of methane per year or decade from agricultural activities, then the 650 million people alive in 1700 should have emitted an amount that was smaller in direct proportion to the reduced size of the population alive at the time (about 11% as much as today). This rough calculation confirms that human agricultural activities should account for the observed rise in atmospheric methane concentration of about 100 parts per billion between 5,000 years ago and AD 1700.

But the observed increase in methane is only part of the story. A complete calculation of the full methane anomaly needs to take into account not just this rise,
but also the amount by which the methane concentration would have fallen (but
did not) during the last 5,000 years if natural processes had remained in control
as they had previously been. Allowing for this additional factor, the full methane
anomaly must be close to 250 parts per billion, considerably more than the observed rise of 100 parts per billion (fig. 8.1).

To explain the full anomaly requires some kind of process that generated
methane in amounts that were disproportionately large compared to the number
of humans alive in 1700. Several of the methane sources noted earlier can seemingly be ruled out because they are tied too closely to population levels (for example, human waste and livestock). I think the most likely source of the "extra"
methane is irrigation for rice farming. Early rice fields were probably far more in-
fested with weeds than their modern-day counterparts, and dying weeds generate
just about as much methane as rice does. I suspect that in the past humans in
Southeast Asia flooded relatively large areas but farmed them in a relatively ineffi-
cient way, yet they were still able to feed the relatively small populations living at
the time. If this is correct, then the amount of methane emitted from those early
weed-infested fields would have been out of proportion to the number of humans
alive.

Just such a trend has occurred in recent decades. From 1950 to 1990, the
amount of rice produced grew twice as fast as the area irrigated (and, presumably,
roughly twice as fast as the methane emitted from that area). The improved rice
yields in the 1900s came about mostly as a result of greater use of fertilizer and
insecticides, as well as genetically engineered improvements in the kinds of rice
grown. This evidence suggests that rice farming in 1950 was only half as efficient
as it was by 1990, so that in effect 1950s' rice farming generated "extra" methane
relative to the food yields. This evidence closes about half of the remaining gap
needed to explain the full methane anomaly of 250 parts per billion.

Humans in the eighteenth and nineteenth centuries lacked advanced agricultural technologies, but they would still have been able to improve rice yields gradually through time by weeding more intensively (as populations and available
labor grew), by selecting the larger rice grains from each year's yield, and by making increasing use of manure from livestock for fertilizer.

In summary, human activities can plausibly explain not just the anomalous rise
in methane concentrations during the last 5,000 years, but also the fact that it did
not fall to lower (natural) levels. If so, humans are responsible for a very large
increase in atmospheric methane (250 parts per billion) over the millennia well
before the industrial era. This early anthropogenic impact amounts to about 70
percent of the range of natural variations that had occurred over the previous

hundreds of thousands of years. In the case of methane, the "anthropogenic era" began about 5,000 years ago.

In the years to come, this hypothesis will need to be weighed against other kinds of information, especially more quantitative kinds of evidence. This task will not be simple. For example, how do you estimate the area devoted to farming irrigated rice several thousand years ago if people have continued to farm the same fields in subsequent millennia, thereby destroying most of the evidence of the earlier impacts? Still, I keep coming back to the central argument that initiated this part of the hypothesis: the fact that the methane trend rose when natural trends predicted a fall. Whatever the specific blend of explanations turns out to be, the reason for this reversal must be humans.

I wonder about that first moment when irrigation was discovered. I imagine someone in a region where the seasonal monsoon rains had once again proven disappointing, looking at a nearby stream or river and then looking back at withering crops. What led to that moment of inspiration? Did a seasonal spurt of river water coming down from distant mountain sources rise to the top of the stream bank, overflow a little, but then start to recede and never reach the fields? Did a fallen tree become lodged in a stream and divert water into a field? Somehow, the idea suddenly came to someone that a small channel dug through that loose soil on the bank of that stream or river would bring water to the failing crops. However it happened, it was a true eureka moment for humankind: free at last from some of nature's whims, and a dependable source of water every summer!

Chapter Nine

TAKING CONTROL OF CO_2

CONVINCED THAT HUMANS had taken control of the atmospheric methane trend by 5,000 years ago, I began to wonder if we might have had a significant effect on carbon dioxide millennia ago. Because CO_2 is a more abundant greenhouse gas than methane, and its effects on climate are generally larger, this question was potentially more important than the methane story, but for a while I resisted pursuing it. One reason for the delay was that CO_2 changes are more difficult to interpret than those of methane. While natural variations in methane are controlled mainly by growth and shrinkage of wetlands at the 22,000-year cycle of orbital change (chapter 5), CO_2 variations occur at all three orbital cycles and are harder to disentangle.

Scientists are also less certain what causes natural CO_2 cycles. The problem is that carbon exists in almost every part of the climate system: in the air as CO_2; in vegetation as grass and trees; in soils as organic carbon; and in the ocean mainly in dissolved chemical form but also in the tissue of organisms. Each of these reservoirs of carbon interacts with the others in different ways and at different rates. Three of the reservoirs exchange carbon relatively rapidly (within days to years): the air, the vegetation, and the upper sunlit layer of the ocean where plankton live. For example, vegetation takes some CO_2 out of the air each spring in the process of photosynthesis but gives it back each autumn when the leaves fall or the grass dies. The amount of CO_2 exchanged between the uppermost ocean layers and the overlying atmosphere depends on factors like ocean temperature and wind strength. The deep ocean holds by far the largest amount of carbon, but it is somewhat isolated from the processes at the surface and interacts with the other reservoirs more slowly. The carbon system is complicated.

One thing we know for certain is the longer history of CO_2 changes. The same ice cores at Vostok Station in Antarctica that yielded the long record of methane also hold a history of CO_2 changes (fig. 9.1). This record has four major cycles, each marked by a slow and erratic drift to lower CO_2 concentrations, and then rapid shifts to higher values every 100,000 years. Smaller cycles at 22,000 and 41,000 years are also present but less obvious.

The concentration of CO_2 ranges from peaks of 280 to 300 parts per million during warm interglacial climates to lows of just under 200 parts per million during major glaciations. These changes are equivalent to transfers of some 200 billion

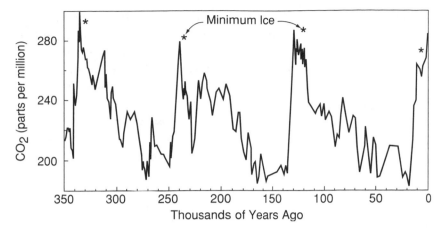

9.1. CO$_2$ concentrations in the atmosphere vary naturally at a cycle of 100,000 years, with peak values occurring a few thousand years before the ice sheets reach minimum (interglacial) size.

tons of CO$_2$. It may seem strange to think of a gas in the atmosphere as weighing billons of tons, but the atmosphere does have "mass" (as in the everyday term "air mass" used by weather forecasters). Consider the weight of a gallon of gasoline you carry in a metal can. Every gallon you burn while driving your car doesn't simply vanish into space; it releases several pounds of carbon to Earth's atmosphere. So if you drive 20,000 miles a year and your car gets 20 miles per gallon, you are personally adding several thousand pounds of gaseous carbon (mostly CO$_2$) to the atmosphere each year.

Climate scientists are still wrestling with the problem of accounting for the 200 billion tons of CO$_2$ that was removed from the atmosphere during major glaciations. Where did it all go? One place it did not go was into vegetation. Huge Northern Hemisphere ice sheets covered regions that had been occupied by forests during warm interglacial climates, so the amount of carbon there in glacial times was much smaller. Also, the glacial world in most places south of the ice sheets was drier and dustier than today, with less vegetation to stabilize the soils from strong winds (chapter 4). As a whole, the continents had less vegetation than they did during interglacial times, at least 500 billion tons less in terms of carbon. So now the challenge is even larger: scientists are faced with explaining what happened to more than 700 billion tons of carbon in glacial times.

The only place left for the carbon to have gone is the ocean, specifically the deep ocean, the largest carbon reservoir of all except for the carbon trapped in sediments and rocks (including coal and oil). How the carbon was transferred to the deep ocean is currently a major area of research. Some of it could have been

sent down by plankton that lived in the surface layers and settled to the sea floor
after they died, taking carbon-rich soft tissue with them. Another way to deliver
more carbon to the deep ocean is for cold water sinking in polar regions to carry
it down in dissolved chemical form. In any case, the experts on this subject know
that CO_2 changes occurred at all three orbital cycles, and they know that the carbon taken from the atmosphere during glaciations went into the deep ocean, but
they don't know for certain which processes were most important.

So, with my expertise lying in other aspects of the climate system, I was reluc-
tant at first to look for possible human impacts on the CO_2 trend in recent millennia. And yet, once again, I noticed a trend that just didn't seem right. If you
look at the CO_2 changes in figure 9.1 and compare the last four times when the
CO_2 values were highest, initially they all appear similar. But if you look more
closely, you will see that they are not.

The first three interglaciations followed the same basic sequence. Peak CO_2
values were reached late in the interval of ice melting and several thousand years
before the times of minimum ice volume (marked by asterisks). All of these CO_2
peaks were coincident with times of maximum summer solar radiation and also
with the largest methane peaks. The fact that these CO_2 peaks occur just before
the ice sheets melted makes sense: the high CO_2 values were helping summer
radiation and high methane concentrations melt the ice sheets. Following these
early peaks, the CO_2 concentrations began steady decreases that lasted for about
15,000 years. Not until some 100,000 years later did the CO_2 values return to
the levels of the previous peaks.

How does this basic pattern compare with the most recent interval of ice melting? A high-resolution CO_2 record spanning the last 11,000 years (fig. 9.2) shows
a trend that looks similar, but only in part: the CO_2 concentration rose rapidly to
a peak of almost 270 parts per million nearly 10,500 years ago. This CO_2 maximum is the same age as the most recent peaks in summer solar radiation and
methane, and it precedes the final melting of the ice (again marked by an asterisk)
by 5,000 years. As a result, this CO_2 maximum 10,500 years ago must be the direct equivalent of those CO_2 peaks that occurred during the earlier interglacial
intervals. After this peak was reached, the CO_2 concentration began to decrease,
just as it had on all three of the earlier interglaciations.

But then, 8,000 years ago, something different happened. Instead of continuing to decrease as expected, the CO_2 concentration began a slow rise that continued until the Industrial Revolution. This behavior may sound familiar. Here
again is the same pattern that occurred with methane: an anomalous recent
increase instead of the decrease expected from earlier (natural) trends.

By the start of the Industrial Revolution, CO_2 values had risen from a low of
about 260 parts per million to between 280 and 285 parts per million. But, as

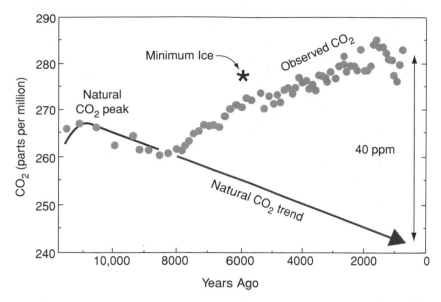

9.2. Natural processes caused the atmospheric CO_2 peak nearly 10,500 years ago and the subsequent decrease until 8,000 years ago, but humans have caused the anomalous CO_2 increase since that time.

was the case with methane, this rise may not be the full anomaly. It is also necessary to try to take into account the amount by which CO_2 would have dropped if natural processes had continued in control during the last 8,000 years. We can do this by examining the previous interglaciations to find the times most analogous to today, checking what the CO_2 values were at those times, and in this way predicting what the natural CO_2 values should be today.

The CO_2 concentrations at those earlier times averaged about 242 parts per million. This value lies along a reasonable projection of the natural CO_2 decrease that occurred between 10,500 and 8,000 years ago before the anomalous reversal in trend (fig. 9.2). If this analysis is correct, then the full CO_2 anomaly by the industrial era would be 40 parts per million, the difference between the 280–285 level reached just before the industrial era and the projected 240–245 natural value. This offset amounts to almost half the range of the natural glacial-interglacial variations (fig. 9.1) and would require the addition of at least 300 billion tons of carbon to the atmosphere during the last 8,000 years. But where could so large an amount of CO_2 have come from?

Two early attempts to explain this anomalous-looking CO_2 increase invoked natural causes. One idea was that the extra CO_2 came from a natural release of carbon from the continents in response to changes in solar radiation caused by

Earth-orbital changes. One part of this idea sounds plausible: we know that the weakening summer monsoons over the last 5,000 years allowed tropical deserts to spread into areas previously occupied by grassland (chapter 5). The net loss of tropical biomass from these changes would have sent CO_2 into the atmosphere. But other evidence refutes this explanation. Global-scale models show that natural carbon losses in some areas over the last few thousand years would have been offset by gains in others. Overall, the loss of biomass carbon is far too small to provide the amount needed to account for the anomalous CO_2 trend.

I concluded that this or any other attempt to explain the rising CO_2 trend since 8,000 years ago based on the natural behavior of the climate system is doomed to fail for a very simple reason. All of the major factors in the climate system (changes in solar radiation, rates of retreat of the ice sheets, rises in sea level, changes in vegetation, and so on) behaved in a similar way throughout the last four intervals of ice melting and then for several thousand years afterward, yet only the current interglaciation shows a CO_2 rise during the early interglacial interval. All of the three prior intervals showed steady drops in CO_2 (fig. 9.1).

As a result, any explanation of the recent CO_2 increase based on natural factors is doomed to fail when tested as an explanation of the CO_2 drops during the three previous interglaciations. With natural explanations apparently eliminated, once again only one explanation seemed to be available: humans must have been the cause of the anomalous CO_2 rise. Somehow, humans had seemingly added 300 billion tons or more of carbon to the atmosphere between 8,000 years ago and the start of the industrial era.

But this conclusion collided head-on with the conventional wisdom. The initial reaction of most climate scientists I spoke to was that the small number of humans on Earth 8,000 years ago could not possibly have taken control of the atmospheric CO_2 trend, especially considering the primitive technologies then at their disposal. This skepticism seemed well justified: How, indeed, could so few humans have had so large an effect? The primary way humans would have released CO_2 to the atmosphere during this interval was by cutting forests. A carbon release of 300 billion tons between 8,000 and 250 years ago would require more than twice as much forest clearance before the Industrial Revolution as has occurred during the 200 years of the industrial era. Modern rates are very high because of rapid clearance of tropical rain forests in South America and Asia. By comparison, the estimated annual clearance rate just two centuries ago was almost ten times smaller than today, and the rates faded away back in time toward trivially small amounts for earlier intervals. From this perspective, it seems preposterous to propose more than twice as much total forest clearance prior to 1750 as afterwards.

But the conventional-wisdom view failed to take into account one key factor—*time*. The average rate of carbon emissions from forest (and other) clearance during the last 200 years has been about 0.75 billion tons per year. At that rate, 200 years of forest clearance would emit a total of 150 billion tons of carbon. But the preceding era of slowly rising CO$_2$ stretches over 7,750 years, an interval 40 times longer. To match the 300-billion-ton carbon emissions target during this earlier era, the average rate of emission would have to have been only about 0.04 billion tons per year, or just 5% of the industrial-era average. With one-twentieth of the rate but an interval 40 times longer, you end up with twice as much in total emissions (divide 40 by 20). As in Aesop's fable, the tortoise (which moves slowly, but starts very early) challenges the hare (faster moving, but late getting started). Look back at figure 1.1. With this simple calculation, the total of 300 billion tons of early carbon emissions began to look more plausible.

Yet this calculation falls well short of a convincing demonstration that humans are responsible for the anomalous CO$_2$ rise. To strengthen the case, it would help to find evidence consistent with two obvious characteristics of the CO$_2$ rise in figure 9.2. First, we need evidence that Stone Age humans began clearing forests at a substantial rate as far back as 8,000 years ago, when the CO$_2$ curve began its rise. Second, we need evidence that the cumulative effects of forest clearance by humans can account for the very large rise in CO$_2$ well before the industrial era (in this case, as early as 2,000 years ago).

Definitive evidence exists that cutting of forests began nearly 8,000 years ago. As agriculture slowly spread from the eastern Mediterranean Fertile Crescent, it eventually reached the forests of southern Europe. As shown in figure 7.2, the distinctive package of cereal grains and vegetable seeds first entered the forested Hungarian Plains nearly 8,000 years ago and then moved north and west into other forested areas. For agriculture to be practiced in forested regions, trees have to be cut to allow sunlight to reach the soil. Much of this clearance was done by slash-and-burn: high-quality flint axes first developed in the Near East nearly 9,000 years ago were used to girdle and kill trees. A year or two later, the fallen debris and the dead trees were burned during the dry season. Crops were then planted in the ash-enriched soil between the dead stumps. In some places, when the soil began to lose its nutrient content a few years later, people packed up and moved elsewhere. In others, they remained and built permanent settlements near their fields.

A wide range of other evidence from Europe points to forest clearance. Along with the distinctive crop grains and seeds, lake sediments contain higher percentages of herb and grass pollen that indicate openings in forests or permanent fields near lakes. Also present are remains of distinctive types of European weeds associated with clearance: plantain, nettles, docks, and sorrels. Increased charcoal in the soils also points to prevalent burning. This evidence from Europe matched the first requirement of the hypothesis perfectly—humans had begun to cut forests in small but significant amounts 8,000 years ago, even though they had only flint axes at their disposal.

Elsewhere, the evidence is less firm, but still suggestive. The first evidence of agriculture in China dates to 9,500 years ago, and by 6,000 years ago pollen sequences in lakes show a noticeable decrease of tree pollen. Although some of this change could reflect moisture loss caused by the weakening summer monsoon, many scientists consider it to be related to increasing human impacts in the valleys of the Yellow and Yangtze rivers. In India, the first evidence of agriculture from the Indus River Valley dates to 9,000 or 8,500 years ago, and expansion of agriculture again required cutting of trees. More studies are needed in both regions to pinpoint the earliest human impacts.

The second requirement—cutting of forests on a much larger scale well before the industrial era—can be tested by examining the start of the historical era, 2,000 years ago. By that time, the world was a very different place compared to 8,000 years ago. In the intervening 6,000 years, human ingenuity had completely altered the practice of agriculture (chapter 7). In place of Stone Age axes and wooden digging sticks were Bronze Age and then Iron Age axes and plows, and domesticated oxen and horses to pull the plows. As a result, evidence of extensive clearance and land disturbance appeared during the Bronze Age in most regions and intensified during the Iron Age.

From the combined efforts of scientists in many disciplines comes a snapshot of world agriculture 2,000 years ago (fig. 9.3). By this time, sophisticated agricultural practices (diverse crops, multiple plantings, and tending of livestock) were in use in China, India, southern and western Europe, and Mediterranean North Africa, as well as the lowlands of Central America, the Amazon Basin, and the highlands surrounding the Peruvian Andes. These regions are all naturally forested, and large-scale agriculture required that those forests be cut.

In such regions, the evidence for increasing human influences on the landscape is plentiful long before the industrial era, but most of this information is not easily converted to specific percentages of forest clearance. By chance, however, the historical literature provided me with one way to make such estimates. In AD 1089 William the Conqueror ordered up the Domesday Survey, a comprehensive analysis of his new domain (England). Included in that survey were two particularly useful numbers: (1) the percentage of land still in forest (rather than in crops or pastures), and (2) the total population. The first number was astonishing: 85 percent of the countryside was deforested, as well as 90 percent of the arable land (elevations below 1,000 m). A full 700 years before the industrial era, Britain was almost entirely deforested, with many of the remaining "forests" protected as hunting preserves for English royalty and nobility. Other evidence indicates that most of this clearance had already occurred by 2,000 years ago. For example, lake

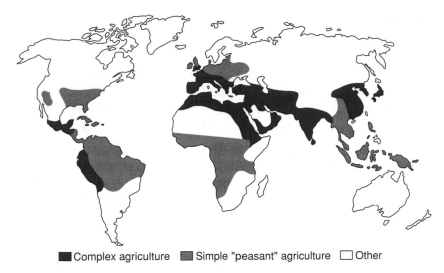

■ Complex agriculture   ■ Simple "peasant" agriculture   □ Other

9.3. By 2,000 years ago, sophisticated forms of agriculture were practiced in naturally forested areas of China, India, southern Europe, and Mediterranean North Africa.

and river sediments recorded large influxes of silt and mud that resulted from destabilization of steeper terrain by deforestation. Here was a firm benchmark that proved massive deforestation far in the past at one location.

The Domesday Survey provided me with an important quantitative link between human population density and the total amount of deforestation. If 1.5 million people required near-total deforestation of a known amount of arable land to make their living as farmers, then every human living in England in 1089 required an average of 0.09 km^2 (9 hectares) of cleared land. Both the clearance and the growth of population had occurred gradually over many millennia, and in effect each new person added to the population base had cleared an average of 0.09 km^2 of "new" land. I think of this number as the "human forest footprint" for Iron Age England.

The Domesday Survey had profound implications for similar changes in other regions. All of the arable land in any region with a population density equal to or larger than that of England in 1089 should have been completely deforested. The earliest reliable census counts in China date from the Han dynasty nearly 2,000 years ago. By that time, China had 57 million citizens at a density some three times that of Domesday-era England. If England was deforested at a much lower density of people, why not China at this higher density? The same conclusion applied to other highly populated areas such as India and Indonesia. This approach suggested that most of the arable land in southern, central, and western Europe should have been deforested by AD 1000.

Gradually I became aware of an enormously rich literature on early deforestation lying scattered among different disciplines. I also discovered a newly published (2003) book by Michael Williams called *Deforesting the Earth*, a superb compilation of the wide-ranging evidence for massive preindustrial deforestation. I learned that much of China was deforested more than 3,000 years ago and that many countries in western Europe began passing laws to protect their remaining wooded areas after AD 700, evidence that the extent of forests had shrunk alarmingly.

Those of us working in the field of climate science tend to think of our research as unusually broad in scope, as indeed it is compared with the kinds of science that focus on smaller-scale processes. To do our job, we need expertise in climatic records from ocean sediments, ice cores, lakes, trees, and corals, and we need to know about the history of changes in the surface and deep ocean, sea ice and ice sheets, the land, vegetation, and air. We also need to be able to fit all these pieces together in order to figure out key cause-and-effect relationships. We can justifiably take some pride in our breadth of knowledge. This also makes studying climate history fun.

Yet here was a case in which at least initially I found myself almost completely ignorant of an entire field of research—the early history of humans and our possible effects on the climate system. I am still filling that gap. Just recently I was made aware of a paper published in 2002 by a group of French scientists who had pointed out that sediments from several regions show increasing amounts of charcoal during the last few millennia that generally match the rise in atmospheric CO_2 observed in the ice cores. Those data also argue for an early human impact on the CO_2 trend.

Although this wide range of evidence argued for large early human releases of CO_2, it did so only in a qualitative way. A more quantitative assessment of the hypothesis was still needed. That assessment started with the Domesday "forest footprint" of 0.9 km²/person for Iron Age England. I then found in the historical ecology literature a similar estimate of the footprint for late Stone Age people in north-central Europe 6,000 years ago. This estimate was derived by calculating the amount of cleared land required for a Stone Age village of 30 people (6 families): the amounts needed for homes and crops, for hayfields and pastures, and for a woodlot cut on a rotating basis. The estimated Stone Age footprint came to 0.03 km² (3 hectares) per person. The tripling of the footprint from 0.03 km²/person 6,000 years ago to 0.09 km²/person in AD 1089 seemed reasonable, given the advent of iron axes and plows and the use of draught animals in the intervening millennium.

Now I had a way to attempt a quantitative estimate of the cumulative amount of carbon emissions by 2,000 years ago. This calculation required three numbers: the human populations (on a regional basis), the amount of forest (in km²) cleared per person, and the amount of carbon emitted by clearing each km^2 of forest. Multiplying these numbers would produce an estimate of the cumulative amount of carbon emitted by total forest clearance as of 2,000 years ago:

$$\text{Population} \times \frac{\text{per-capita forest}}{\text{clearance}} \times \frac{\text{tons of carbon emitted}}{\text{by clearance}} = \frac{\text{total carbon}}{\text{emitted}}$$

$$\text{(people)} \qquad \text{(km}^2\text{/person)} \qquad \text{(C/km}^2\text{)} \qquad = \qquad \text{C}$$

The number of people alive 2,000 years ago was available from historical population estimates. Census data were available from Europe and China, where head counts were made primarily for tax collection purposes (yes, even 2,000 years ago!). Populations were less well known in areas like India, Indochina, and the Americas. In all, some 200 million people were alive 2,000 years ago.

I grouped these 200 million people according to whether they lived in naturally forested regions or nonforested areas (deserts and steppes). Roughly 10 percent lived in deserts and steppes and probably had no significant effect on carbon emissions. The other 90 percent lived in naturally forested regions, where practicing agriculture meant cutting forests. The people living in naturally forested regions were then divided into Iron Age cultures (all of Eurasia and Mediterranean North Africa) with a forest footprint of 0.09 km^2/person and Stone Age cultures (the Americas and sub-Saharan Africa) with the smaller 0.03 km^2/person footprint. At this time, by far the most people (130 million) lived in the Iron Age cultures of southern Eurasia (China, India, and Europe).

The final number needed for the calculation—the amount of carbon emitted by clearing 1 km^2 of forest—came from the field of ecology. This number ranged from 1,000 to almost 3,000 tons of carbon per km^2, depending on the type of forest (more from dense tropical rain forests, less from forests in areas with seasonally dry or cold climates). Each region was partitioned into the type of natural forest that ecologists estimated would exist in the absence of human influences.

The total figure I arrived at was large: 200 billion tons of carbon or more could have been released to the atmosphere from clearance by 2,000 years ago. This value was well along the way toward the more than 300 billion tons estimated for all preindustrial carbon releases. It seemed likely that the second requirement of the hypothesis might be met: forest clearance might well be able to explain the growing size of the CO$_2$ anomaly over the last 8,000 years.

I also discovered other kinds of human activities that would have released carbon long ago. As forest fuels began to disappear from some regions in northern Europe, people began unearthing and burning deposits of peat for cooking and for heat. A crude attempt can be made to calculate how much peat carbon might have been added to the atmosphere in preindustrial times: If an average of 5 million households burned 10 half-kilogram (~1-pound) bricks of peat every day of the year over a time interval of 2,000 years, some 10 billion tons of carbon could have been added to the atmosphere during that time. In addition, the Chinese have been mining and burning coal for the 3,000 years or more since they deforested the arable lands in the north-central parts of their country. A calculation similar to that for the burning of peat suggests the possibility of another 20 billion tons of carbon emissions from China over that interval.

Although these findings seemed promising, they certainly did not settle the question of the size of the human impact on the atmospheric CO_2 trend over the last 8,000 years. I suspect the debate over this issue will go on for years and maybe decades. But I keep coming back to that one central fact I noticed at the start of my investigations: the CO_2 concentration has risen during recent millennia, yet it had always fallen during the most similar intervals of previous interglaciations. It went up when it should have gone down. Humans are the most obvious explanation for this wrong-way trend.

In a way, this hypothesis challenges many of our basic instincts about what is or is not natural. Think of a beautiful Mediterranean Sea coast with blue-green water and grassy hillsides grazed by goats tended by herders. As beautiful as that scene is, it is anything but "natural." If those goats were kept out and nature were allowed to reassert control for a century or more, some of these regions would once again be covered with rich Mediterranean forests. Other hillsides have lost so much of their soil cover and have so few nutrients left in the soils that remain that they cannot revert to their naturally forested state.

As for those charming Mediterranean seaports tucked into the coastline between the steep hills, many of the ports near river mouths were located farther inland 2,000 years ago. As early deforestation stripped the terrain of its natural cover, the steeper slopes were no longer able to retain the soil when heavy rains fell. Eroded silts and muds began to clog the ports, forcing relocation of coastal towns seaward to catch up with the advancing shorelines. Similar stories apply across southern Asia. By 2,000 years ago, that part of the world had been heavily transformed by humans, and nature was no longer in control of the landscape. Nor was it in control of the atmospheric CO_2 trend. Humans had taken control.

Chapter Ten

HAVE WE DELAYED A GLACIATION?

THE EVIDENCE SEEMED CLEAR: human activities linked to farming had taken control of the trends of two major greenhouse gases thousands of years ago, forcing their concentrations to rise when nature would have driven them lower. The net impact of humans through time (fig. 10.1) was a long slow rise in greenhouse-gas concentrations prior to the industrial era, and then much more rapid increases during the last 200 years of industrialization.

Scientists use a convenient standard to evaluate the climatic impact of greenhouse gases: the amount by which Earth's climate would warm or cool if the CO_2 concentration were either doubled or halved from the preindustrial concentration of 280 parts per million. This number, called the $2 \times CO_2$ ("doubled CO_2") sensitivity of the climate system, averages 2.5°C (4.5°F) for the planet as a whole. Changes in concentrations of other greenhouse gases can be converted to a form that is expressed as an equivalent change in CO_2. The exact size of this $2 \times CO_2$ sensitivity is uncertain; it lies somewhere between 1.5°C and 4.5°C, with 2.5°C the current best estimate.

For a climate sensitivity of 2.5°C, the slow buildup of CO_2 and methane caused by human activities before the start of the industrial era would have warmed global climate by about 0.8°C (just under 1.5°F). This number sounds small but is far from trivial; it is roughly 15% as large as the cooling that occurred at the last glacial maximum, the time when most of northern North America and northern Europe was buried under ice sheets.

Climate scientists have long viewed the last 8,000 years as a time of naturally stable climate, a brief interlude between the previous glaciation and the next one. But the story presented here suggests this warm and stable climate of the last 8,000 years may have been an accident. It may actually reflect a coincidental near-balance between a natural cooling that should have begun and an offsetting warming effect caused by humans. If this new view is correct, the very climate in which human civilizations formed was in part determined by human farming activities. Even thousands of years ago, we were becoming a force in the climate system.

The high latitudes are the most sensitive part of the climate system. Temperature changes there are two to three times the size of the global-average value. The main reason for this larger responsiveness is the presence of snow and ice and their ability to reflect the Sun's radiation. If for some reason the planet cools,

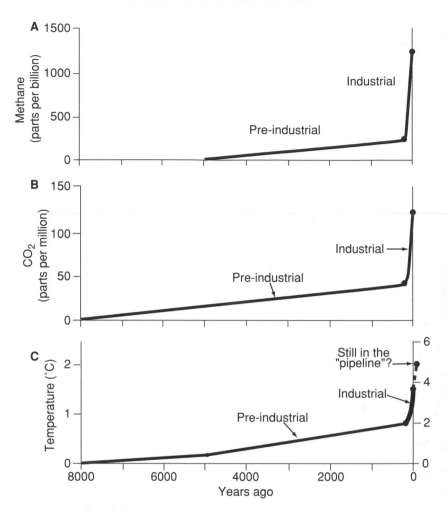

10.1. Humans have had a large impact on atmospheric methane concentrations (A), CO_2 (B), and estimated global and high-latitude temperature (C left, right) during the last several thousand years.

snow cover on land and sea-ice cover on the ocean will expand, especially at frigid latitudes near and poleward of the Arctic Circle. The greater expanses of these bright white surfaces will reflect much more incoming solar radiation than the dark-green land surfaces and deep-blue ocean they replace. Land and water are heated by the solar radiation they absorb, but snow and ice reflect almost all incoming radiation back to space. In this way, increased expanses of snow and sea ice in the Arctic amplify and deepen any cooling by a "positive feedback." The same positive-feedback process works when climate warms. Warming causes snow and sea ice to retreat, which exposes more of the darker land and ocean, which absorb more solar radiation, which makes the Arctic warmer by an extra amount beyond the initial warming.

Allowing for this poleward amplification of temperature changes, the 0.8°C global-mean warming caused by preindustrial human inputs of greenhouse gases should have been about 2°C (3.6°F) in Arctic regions (fig. 10.1C). This warming is sizeable, but it has been masked by a somewhat larger natural cooling driven by changes in Earth's orbit. As noted in chapter 4, the intensity of summer solar radiation at high latitudes has decreased by about 8 percent since 11,000 years ago. This reduction in radiation has slowly nudged high-latitude climate toward colder summers, even though human-generated greenhouse gases have been pushing in the opposite direction.

Several natural climatic sensors have registered a slow summer cooling trend in the Arctic over the last several thousand years. One is the slow southward retreat of the northern boundary of spruce forest, with frigid Arctic tundra advancing southward and replacing trees. Another example comes from sediments in the Norwegian Sea, where species of plankton that prefer to live under and around sea ice have become more abundant, indicating more prevalent sea ice in recent millennia. Still another line of evidence is recorded in small Arctic ice caps, which contain layers showing significant melting during summers 10,000 years ago, but few such layers in recent millennia.

The northeastern Canadian Arctic is especially interesting to climate scientists because of its role in the long history of ice-age cycles over the last 2.75 million years. This frigid region was the place where the last remnants of the most recent ice sheet melted (see fig. 4.1). A gigantic dome of ice had extended halfway to the equator until 16,000 years ago but then had gradually retreated to the northeast during the next 10,000 years and ended up as small remnants in northeast Canada by 7,000 to 6,000 years ago. In this region today, small and rapidly melting remains of ice caps still lie perched on high terrain from Baffin Island at 65°N north to Ellesmere Island at 83°N, just about the same range of latitudes as the Greenland ice sheet on the opposite side of the Labrador Sea.

Many scientists think that those many dozens of glacial cycles that have occurred during the last 2.75 million years began in this very region. Milankovitch had suggested that snowfields grow into small ice caps and then large ice sheets when summer solar radiation grows too weak (and summers too cold) to melt the snow. Summers are cold in this area for two reasons: in part because it lies so far north, and in part because the northeast coast of Canada is a slightly elevated block. Given the fact that the snow cover normally melts away for just a couple of months in a typical modern summer, this region is obviously close to a state of glaciation right now. This made me wonder: What would it take to push it over that edge? More specifically: Would this region now be glaciated had it not been for the warming from those greenhouse-gas releases caused by farming?

Climate scientist Larry Williams explored this issue in a climate-modeling analysis of this region in the 1970s. He found that a cooling of about 1.5°C (2.7°F) relative to 1970s' temperature levels should be enough to cause permanent snow cover and glaciers to form on the higher terrain along the edges of the Labrador Sea. In effect, the small ice caps that now exist there would have expanded into somewhat larger masses of ice, although still confined to this far northeastern corner of Canada. His study further suggested that an additional cooling of 1–2°C (1.8–3.6°F) would be enough to cause permanent snowfields to form over a larger area beyond Baffin Island.

This modeling study from almost 30 years ago ties in directly with the greenhouse-gas history shown in figure 10.1. Williams's estimates can be used to peel away the warming effects caused by human inputs of greenhouse gases and predict whether or not large glaciers would now exist. Williams estimated the industrial-era warming in this region over the 100 years prior to his study at 1.5°C. He noted that if this recent warming were removed, the highest terrain of Baffin Island should be brought right to the edge of glaciation. In confirmation of this conclusion, the ice caps that now exist on Baffin Island are thought to have formed in the cooler climate just prior to the industrial-era warming of the last century. Because of the subsequent warming, these ice caps are rapidly melting, and many are likely to disappear in the next few decades.

The second greenhouse-gas warming that can be peeled away is the 2°C preindustrial warming caused by humans and the expansion of early agriculture (fig. 10.1C). With this earlier warming effect removed, Williams's results suggest that a broader area of high terrain in northeast Canada would have been brought to a state of permanent snow cover and thus glaciation.

The implication of these model results, combined with the new evidence for early human generation of greenhouse gases, was startling. Part of northeast Canada might well have been glaciated before the industrial era had not our early greenhouse-gas emissions occurred, and ice would exist in the region today were it not for the combined effects of the preindustrial and industrial-era gas emissions. It seemed that human greenhouse gases have stopped a glaciation!

This startling conclusion triggered another recollection from earlier climate studies. In 1980 geologist John Imbrie, a central figure in confirming the Milankovitch ice-age theory (chapter 4), had enlisted his mathematician son to try to devise a simple method to convert the well-known changes in solar radiation through time into an estimate of the history of global ice volume. The logical basis of this exercise was straightforward. For the last several thousand years, the ice sheets have regularly grown and melted at the three major orbital cycles of

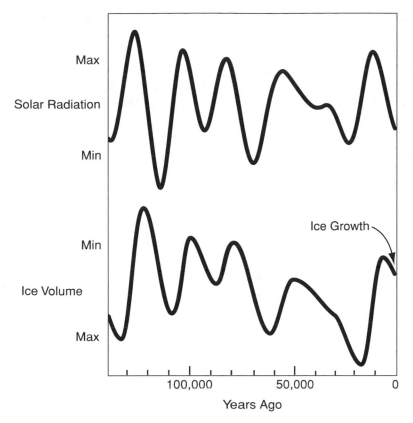

10.2. Several models of the ice-sheet response to July solar radiation changes in the Northern Hemisphere predict that ice sheets should have begun growing several thousand years ago.

100,000, 41,000, and 22,000 years in response to changes in solar radiation (chapter 4). Given the evidence for this long-term relationship, why not try to link the two in a mathematical formula?

For the "target signal" in their exercise, the Imbries used the marine record of oxygen isotopes, which provided an estimate of the size of the ice sheets through the last several hundred thousand years (chapter 4). However, even though the same orbital cycles are present in both the solar radiation and the ice-volume signals, linking the two is not as easy as it might sound. The main problem is that the 100,000-year signal is much stronger in the ice-volume changes than it is in the solar radiation changes, and the reason for this mismatch is not well understood.

The result of the modeling exercise is shown in figure 10.2. When the Imbries published these results, they noted that their model matched the past behavior of ice sheets well enough to conclude that natural orbital variations should carry Earth's climate into the next glaciation within the next few millennia (ignoring current and future human impacts on greenhouse gases). But looking over their results from a fresh perspective, I realized that their model had revealed something they had ignored. Their simulation showed that the current interglaciation should have peaked about 6,000 to 5,000 years ago and that the climate system should have been slowly drifting back into a new glaciation since then. The Imbries would have had valid reasons for ignoring this part of their model's simulation, including complications from temperature overprints on the ice-volume target signal they were trying to match.

Probably the most important reason for ignoring the model simulation of the last 5,000 years would have been the simple fact that no glaciation has developed. This would have made it appear that the model simulation failed during this interval. But now I was looking at this 20-year-old paper from a different viewpoint: Could the model simulation for the last 5,000 years actually be correct? Did this seeming "failure" result from humans having overridden the natural behavior of the climate system by introducing greenhouse gases and stopping the very glaciation the model had simulated? From this perspective, the Imbrie model provided me with an additional line of support for the idea that a new glaciation is actually overdue.

The Imbries' results are a direct outgrowth of Milankovitch's original ice-age theory. Long ago he suggested that ice sheets are so sluggish in their response that changes in their size lag an average of about 5,000 years behind the changes in summer radiation that cause them to grow or melt. The last maximum in summer radiation at the latitudes of the ice sheets occurred 10,000 years ago, after which the solar radiation levels have continually fallen. Allowing for the 5,000-year lag that Milankovitch invoked, the ice sheets should have begun growing about 5,000 years ago, just when the Imbrie model suggests.

I also reexamined simulations from other climate models that had been developed in the last two decades to simulate ice-sheet trends in the past and then project the trends into the future using known changes in Earth's orbit. Some (but not all) of these models had also predicted that at least a small amount of ice should have started forming sometime within the last several thousand years. And yet the scientific community had ignored the basic discrepancy between the model simulations of renewed glaciation and the fact that no ice sheets have formed.

Piecing this information together, I came up with the scenario in figure 10.3. Natural climate was warm 8,000 years ago because of strong solar radiation in summer and high natural levels of greenhouse gases. Since that time, the summer solar radiation level has been falling in response to natural (orbital) changes and causing a natural cooling. Within the last few thousand years, this cooling would

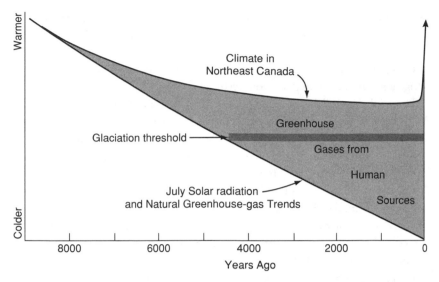

10.3. A natural cooling trend in the Northern Hemisphere should have passed the threshold for initiating a new glaciation several thousand years ago, but greenhouse gases added by humans kept climate warm enough to avoid the start of a new ice age.

have reached the threshold at which glaciation became possible, but humans had begun adding greenhouse gases to the atmosphere in amounts sufficient to keep climate warm enough to avoid glaciation. Overall, polar climates did cool during the last 5,000 years because of the solar radiation changes, but ice sheets failed to form because of the human additions of greenhouse gases. During the last two centuries, industrial-era emissions have driven greenhouse gases farther above the temperatures at which glaciation is possible.

I knew that this hypothesis—that an ice sheet of some size should now exist in northeastern Canada—would be provocative and controversial, and I turned to a long-time friend, John Kutzbach, and his colleague Steve Vavrus to try to test the idea a different way. We ran an experiment using the same kind of model that Kutzbach had relied on to test his orbital-monsoon hypothesis (chapter 5), a model that reproduces in three dimensions Earth's climatic response to any kind of nudge away from its present state. The present-day climate is the baseline state in the model, but it can be altered for such experiments.

Our experiment was designed to answer two simple questions: How much cooler would Earth be today if no greenhouse gases of human origin existed in the atmosphere? And, more specifically, would it be cold enough to support a new ice sheet in Canada, or anywhere else for that matter? In our experiment, we made just one change in the baseline state, removing all human-generated greenhouse gases from the model's atmosphere. Out came the CO_2 and methane generated by humans prior to the industrial era (based on the analysis described in chapters 8 and 9), and out came the additional gases generated during the industrial era, including not just CO_2 and methane but also chlorofluorocarbons (CFCs) (from refrigerants) and nitrogen gases from fertilizers and other sources.

The resulting model simulation produced a global-mean cooling of just under 2°C. This is a very large number: on average, Earth was only a little more than 5°C colder during the last glacial maximum. Slightly under half of the simulated 2°C cooling in our experiment was caused by removing the preindustrial gases and a little over half by removing the industrial-era gases. At higher latitudes of both hemispheres, the amount of cooling was larger, especially in the winter season when model-simulated changes in snow and sea-ice extent amplify the cooling. The largest cooling over any continent was centered on northern Hudson Bay in east-central North America, with a mean-annual decrease of 3–4°C and a winter cooling of 5–7°C (fig. 10.4).

Such a deep cooling would seemingly be favorable for the growth of ice sheets, although the kind of model used in this experiment can give us only a partial answer to whether or not ice actually grew. The model simulates changes in temperature, precipitation (rain or snow), and sea-ice extent and thickness across the entire surface of the planet, as well as properties like wind strength and atmospheric pressure in the overlying atmosphere. But simulating so many variables in so many regions comes at a cost. The model is so complex and expensive to run that only a few decades of Earth's climatic response can be simulated in any one experiment. As a result, very slow processes that develop over many millennia are beyond the scope of this kind of model, including the gradual buildup and melt-back of ice sheets.

Still, the model gives an indication of whether or not ice is likely to have formed in any particular region. It simulates the thickness of snow cover (and sea ice) throughout the year, balancing the accumulation of snow and ice in autumn and winter against the melting in spring and summer. The critical issue is whether or not any snow cover or sea ice persists through the entire summer. If it does, the next winter's snow can add to the thickness, and then the next and the next. In time these snowfields will coalesce and turn to solid ice, marking the beginning of glaciation.

The model results showed just one region on Earth entering this state of "glacial inception"—Baffin Island, the location of Williams's modeling study. Today, Baffin Island is snow-free for an average of 1 to 2 months each summer, both in the model baseline simulation and in reality. But for the simulation in which we removed human-generated greenhouse gases, snow now persisted year-round in a few areas of higher terrain along the high spine of Baffin Island (fig. 10.4). The

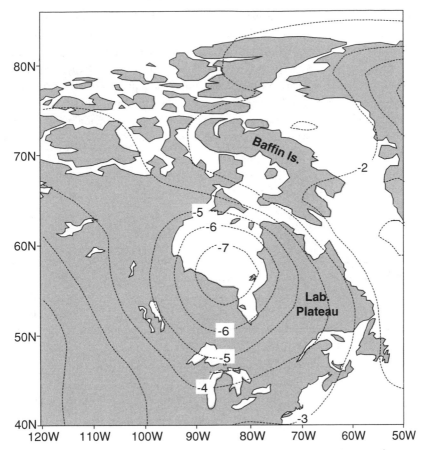

10.4. When greenhouse gases of human origin are removed from the atmosphere in a climate-model run, North America becomes much cooler in winter, and snow persists year-round on Baffin Island and for 11 months across high terrain in Labrador.

simulation also showed that that a second area just east of Hudson Bay moved much closer to a state of incipient glaciation. In the Labrador region of eastern Canada, a large plateaulike feature rises to an elevation of 600 meters (2,000 feet). In the baseline run, this plateau had been snow-free for 2 or more months each summer, but in the run with lower greenhouse gases it was snow-free only during the month of August.

The fact that these two regions, and *only* these two, were in or close to a state of incipient glaciation is highly suggestive if you look back at the map of the melting North American ice sheet shown in figure 4.1. Baffin Island was the last area from which the great ice sheet melted, and it is the first region the model simulates as poised to enter a new glacial cycle. The Labrador Plateau was the next-to-last region to lose its glacial ice cover, and the model simulation shows it as the next area likely to return to a glacial state. In addition, the model results with lower greenhouse-gas levels show sea ice lasting one month longer in parts of Hudson Bay and disappearing for just the month of September, again conditions very close to those needed for glaciation. This simulation provided confirmation for the hypothesis that at least a small part of northeastern Canada would be in a state of incipient glaciation were it not for humans and our early greenhouse-gas emissions.

Still, I was initially disappointed with the results from this simulation. I told my colleagues that I had hoped we might hit a home run with this experiment, with clear indications of incipient glaciation across a broad area, but the localized evidence of a few regions with snow persisting through summer felt more like beating out an infield single.

Yet it was worth keeping in mind that several important processes were absent from the model we used for this experiment. At high northern latitudes, climatically driven changes in vegetation provide an important positive feedback that can amplify the size of initial changes in climate. One important change is linked to shifts in the boundary between the northern limit of conifer forests and the southern limit of treeless tundra. When cooling occurs along this boundary, the forest retreats southward and is replaced by tundra. In such areas, the dark-green, heat-absorbing forest canopy is replaced by bright white, snow-covered tundra (mostly grass and low-lying shrubs). The advancing tundra reflects more solar radiation than the forest it replaces, and so the area cools by an additional amount. This added cooling in turn favors longer persistence of a thicker snow pack into and through the summer months. Our first experiment did not include this positive feedback. But results from other experiments indicate that the large size of the cooling shown in figure 10.4 would push tundra south of its present-day limits, further cool local climate, and increase the extent and persistence of snow and sea-ice cover. Future runs that include these and other feedbacks will likely strengthen our conclusion that at least a small glaciation is overdue in northeastern Canada.

The question of how big an ice sheet might exist in Canada today if nature were still in control is difficult to answer. Baffin Island is the most likely place to have been glaciated, and it is similar in size and dimensions to the combined size of California and Oregon, spanning a similar 10° range of latitude. The high plateau in Labrador that moved close to a glaciated state in the model experiment is about the same size as New England and New York State combined. If future experiments show that both areas would have begun to accumulate ice, the extent of that ice sheet would be much smaller than the enormous glacial-maximum ice sheets, and yet larger than the size of the present-day ice sheet on Greenland. Small and yet big, depending on how you look at it. And in any case, ice was present on this continent and slowly growing toward the south and west.

This reinterpretation of the basic course of climate change during the last several millennia sheds a revealing light on one aspect of the current policy debate about global warming. The confirmation in the 1970s of Milankovitch's theory that gradual orbital variations controlled the growth and decay of ice sheets was one of the great success stories of climate science (chapter 4). The 1980 paper by the Imbries extended these findings, showing that the next glaciation was "imminent," that is, due within no more than a millennium or two. Unfortunately, a few scientists at the time used these results to jump to an erroneous short-term conclusion: they inferred that the brief leveling out and slight downturn in global temperature under way during the 1960s and 1970s might be the onset of a new ice age.

By the 1980s, decades of direct measurements of rising CO_2 concentrations in the atmosphere had convinced most climate scientists that the rapid CO_2 increase would be a much larger factor in climate change in the immediate future than the much slower orbital changes. As a result, concerns about near-term warming, rather than longer-term cooling, came to dominate deliberations that by then were beginning to receive unprecedented attention from the media. At times, spokespersons for environmental organizations made exaggerated, alarmist statements about the possible harmful effects of future warming.

The 1990s saw a backlash against these alarmist statements, with doubts expressed that the scientific community had a good grip on the complexities of this issue. The switch from predictions of a future glaciation to warnings of a future heat wave was (and still is) cited as an example of scientific incompetence.

The results summarized here bring a wider perspective to this issue. Most scientists who drew on the newly confirmed orbital theory in the 1970s to infer that a glaciation was imminent were clearly doing so in the long-term context of the slow orbital cycles. They saw a new glaciation as being "imminent" in the sense of being due within a few centuries or millennia, not tomorrow. The reinterpretation in this chapter suggests that most of those scientists were actually understating the case, rather than being alarmists. The findings described here indicate that Earth should have undergone a large natural cooling during the last several thousand years, and that at least a small glaciation would have begun several millennia ago had it not been for greenhouse-gas releases from early human activities. The next glaciation is not "imminent"; it is overdue.

On the other hand, the few scientists who jumped to the premature conclusion that the minor climatic cooling during the 1960s and 1970s was a harbinger of glaciation deserve the criticism they have received. They were simply wrong.

Chapter Eleven

CHALLENGES AND RESPONSES

MOST NEW SCIENTIFIC IDEAS follow a typical sequence. After the "thesis" (publication of the new hypothesis) comes the "antithesis" (evaluation and criticism by the scientific community), and later, for those hypotheses that survive close scrutiny, the "synthesis" (refining or reshaping of the hypothesis into a form that addresses the criticisms and satisfies a wider range of observations). At this point, usually years later, the hypothesis may come to be called a "theory."

In the case of my hypothesis, the thesis stage occurred the first week of December 2003, when my paper on early human impacts on climate was published in the journal *Climatic Change* and when I gave a lecture at the annual meeting of the American Geophysical Union in San Francisco. Almost 10,000 scientists attended that meeting, and more than 800 of them were in the lecture hall or spilling out into the nearby hallways on that Tuesday evening. The initial impact of the hypothesis seemed very positive, although this "honeymoon phase" would prove to be brief. In the next few days, reports of the talk and article appeared in major media outlets, including the *New York Times*, London *Times*, *Economist*, Associated Press, and BBC radio. Dozens of people came up to me at the meeting over the next few days voicing favorable opinions, often with obvious enthusiasm. Near the end of the meeting, a climate-modeler friend and self-described skeptic told me that he had talked to other skeptics in related fields and that all of them had the same reaction to my talk: "Damn, why didn't I see that?"

This initially favorable reception lingered into the winter of 2004, with feature articles in two prestigious scientific journals, *Science* and *Nature*. Both articles had the tone of cautious objectivity appropriate to commentary on a newly published idea, but both also cited noticeably positive comments from prominent scientists, and little by way of direct criticism. Later in the winter I received word that a joint proposal I had submitted to the National Science Foundation with colleagues at the University of Wisconsin to explore this hypothesis had been ranked highly by external reviewers and a panel of experts, and that it would be funded.

With all this good news, I began to wonder if the hypothesis might be accepted more quickly than I had expected. In the spring of 2004, however, two published papers raised serious challenges that needed to be addressed. The standard antithesis phase of the science process had begun. Those challenges, and my responses to them, are the subject of this chapter.

One challenge was that I had not chosen the optimal part of Earth's climatic record as an analogy for the natural trend that climate would have taken during the last few thousand years in the absence of human impacts. As explained in chapters 8 and 9, I had compared greenhouse-gas trends in recent millennia with those during the early stages of the last three interglaciations because those were also times (like today) when ice sheets had melted from Northern Hemisphere continents, and when changes in solar radiation were moving in the same direction as now. During those intervals, the natural concentrations of methane and CO_2 had dropped, whereas levels of both gases had risen in the last few thousand years (fig. 11.1). I had concluded that only human activities can explain the rise in gas levels in recent millennia compared to the natural drops in earlier times.

The criticism of my method was that I should have looked farther back in the sequence of ice-age cycles at evidence from an interglaciation that occurred 400,000 years ago. Solar radiation trends at that time were the closest available analog to the trends during the last few thousand years, whereas the radiation trends during the times I had used for comparison were larger in magnitude, even though moving in the same direction.

This criticism was valid; the earlier interglaciation 400,000 years ago is indeed the closest analog to the recent radiation trends. Figure 11.2 shows the summer solar radiation trends at 65°N for the earlier interval and for the last few millennia (and several thousand years into the future). The amplitudes of the two trends differ a little: the amount of radiation during the earlier interglaciation is a bit lower than that during the last 10,000 years, but the differences compared to the recent trends are a factor of 2 or more smaller than during the intervals I had used in my paper.

I had previously avoided attempting a comparison with this earlier interval because the one long ice-core record available from Antarctica bottomed out in the very interglacial interval of interest, and I was concerned that it did not quite reach the level that was the best analog to the last few thousand years. But now that I was hearing this criticism (from several directions), it was clear that I needed to take a closer look at that older part of the signal.

As it turned out, the ice core did reach far enough back in time. Three independent methods of estimating the age of the deeper ice converged on the same time scale, and the interval most similar to today was indeed present.

Here was a moment of truth: the answer would come quickly and would likely be decisive. If the greenhouse-gas concentrations in the ice during this earlier interglacial analog dropped more or less as I had predicted (fig. 11.1), the implication would be that the gas increases of the last few millennia are indeed anomalous, and thus the result of human activities. But if the earlier trends followed

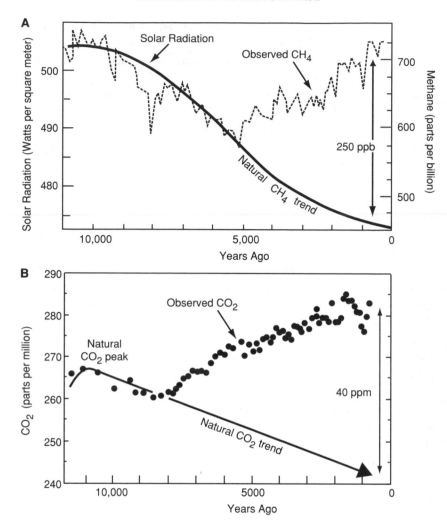

11.1. Concentrations of methane (A) and CO_2 (B) should have fallen during the last several thousand years but instead rose because of human activity.

more or less the same rising trend as has occurred during the last few millennia, the implication would be that the recent trends are natural, and my hypothesis would be invalid.

The answer was clear, and it was decisive: concentrations of both methane and CO_2 fell during the earlier interglaciation to levels very near those I had proposed (fig. 11.3). The methane value fell to a level slightly lower than predicted, while the CO_2 value fell to a value slightly higher, but in each case both the fundamental downward direction of the trends and the levels reached vindicated the

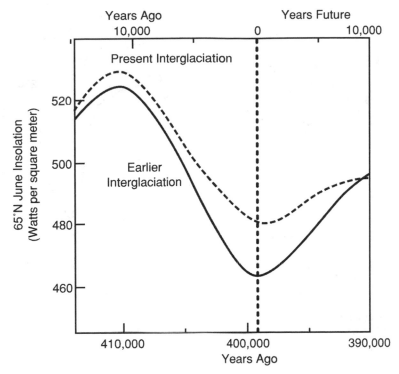

11.2. Changes in summer solar radiation nearly 400,000 years ago are the closest analog to July radiation changes during the last several millennia.

hypothesis. The greenhouse-gas increases during the last few thousand years have indeed been anomalous compared to the natural behavior of the climate system. And if these increases were not natural, they must have originated from human activities.

Yet at first this evidence appeared to conflict with other analyses of this same interglaciation 400,000 years ago. Research on marine sediments had indicated that the interval of interglacial warmth at that time had lasted for an unusually long time, and results from a newly drilled Antarctic ice core had confirmed that conclusion. These two lines of evidence made a convincing argument that the earlier interglaciation had lasted much longer than the present interglaciation has to date.

Based on this evidence, some scientists had concluded that the current interglaciation still has thousands of years to run until the "natural" shift toward glacial conditions begins. This interpretation seemed reasonable, but it directly conflicted with my hypothesis that a significant climatic cooling and new glaciation are now overdue. The next glaciation can't be both overdue and also far in the future. One of these interpretations must be incorrect, but which one?

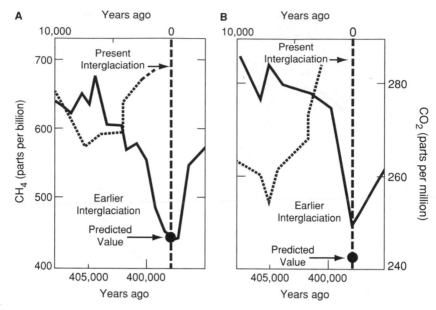

11.3. During the interglaciation nearly 400,000 years ago, methane and CO_2 concentrations fell to natural values much lower than those reached during recent millennia.

A closer look at the evidence pointed me to the answer. During the same interval in the Vostok ice core nearly 400,000 years ago when the CO_2 and methane values dropped (fig. 11.3), Antarctic temperatures had also plummeted from relatively warm interglacial levels to the colder values typical of glacial conditions (much colder than the temperatures in that already frigid region today). This evidence meant that the warmer interglacial conditions in Antarctica had ended at a time that was fully consistent with my hypothesis—during the interval just after 400,000 years ago when solar radiation changes were most similar to those today.

This additional information meant that the long interval of interglacial warmth 400,000 years ago must have begun earlier (in a relative sense) than the current interglaciation. An earlier start to the warm interval would permit that interglaciation to last for a long time and yet come to an end at the time when solar radiation trends were most analogous to the present day. The apparent conflict with my hypothesis was not a real one.

The evidence from marine sediments presented a similar challenge. As in Antarctica, the North Atlantic Ocean had warmed early at the start of the older interglacial interval 400,000 years ago, and it had also remained warm for an unusually long time. But in this case, the ocean south of Iceland had stayed warm through the interval when my hypothesis predicted new ice growth in the Northern

Hemisphere. Because a warm ocean seemed to imply an interglacial world, some scientists had concluded that the world must have remained free of ice sheets through an interval when my hypothesis predicted they should be growing.

I rejected the interpretation that this ocean warmth meant that ice sheets were not growing. More than 25 years ago, a colleague, Andrew McIntyre, and I had shown that much of the subpolar North Atlantic Ocean south of Iceland stays warm at times when ice sheets are growing on nearby continents. Why it does this is not clear, but oceans are not simply passive responders to climate change. Each region has its own characteristic dynamic response that can allow it to react to local changes in winds and other factors. In the North Atlantic, that local response for some reason tends to create "lagging ocean warmth" during ice growth.

Other important evidence comes from the Atlantic Ocean north of Iceland. A study of ocean sediments in that area found that icebergs began to drop large quantities of coarse debris to the sea floor just after 400,000 years ago, after an interval free of such deposition. To generate icebergs that reached the ocean and dropped debris, ice sheets must already have been growing for several thousand years. If the estimated timing of this iceberg influx is correct, ice sheets must have been growing at the same time that the North Atlantic south of Iceland was still warm, in agreement with the hypothesis.

The amount of ice at this time was probably not very large, perhaps no more than 10 percent of the volume of ice typically present when each 100,000-year cycle of ice growth culminates in glacial-maximum conditions (chapter 4), and possibly less. This modest amount of ice apparently was not enough to make its presence felt by sending cold winds into the more temperate latitudes of the North Atlantic. Yet it would likely have been a larger volume of ice than the present-day Greenland ice sheet.

In summary, the evidence for a long interval of interglacial warmth in the North Atlantic did not disprove my hypothesis, and other evidence from the marine record supported it. During the earlier interglaciation 400,000 years ago, at a time when solar radiation trends were most similar to those during recent millennia, a significant natural cooling had occurred in many regions, and a new glaciation (probably small in scale) had begun in the Northern Hemisphere. In contrast, during the last few thousand years, few regions on Earth have cooled, and no new ice has appeared in northern lands. The only plausible reason for the differences in climatic response during these two intervals is that emissions of greenhouse gases by human activities averted most of a natural cooling that would otherwise have occurred and thereby prevented a glaciation from getting underway. A glaciation is now overdue, and we are the reason.

Another major challenge to my hypothesis appeared in a paper whose authors claimed that humans could not possibly have cleared and burned enough forest vegetation in preindustrial times to account for the 40 parts per million size I had estimated for the CO_2 anomaly (fig. 11.1). This criticism also appeared to have considerable merit.

My calculations of forest clearance had suggested that more than 200 billion tons of carbon could have been emitted by human activities, an estimate that seemed reasonably close to the amount needed to explain the 40-ppm CO_2 anomaly. But now a model simulation had been run that required a much larger amount of carbon to explain the CO_2 anomaly—somewhere in the range of 550 to 700 billion tons. Because humans could not possibly have released that much carbon into the atmosphere in the last few thousand years, the authors of this paper concluded that human activities could not account for a 40-ppm CO_2 anomaly. If so, my hypothesis must be flawed.

Here was another impasse—two lines of evidence that again seemed to be in direct conflict. On the one hand, my new investigations had seemingly confirmed that the CO_2 anomaly that developed in recent millennia could only be the result of a "new" process that had not been present during the four previous interglaciations, and forest clearance for agriculture was the most obvious candidate. Yet the model results indicated that human biomass burning could not account for the 40-ppm anomaly. How could these seemingly conflicting lines of evidence be resolved?

I looked again at my original definition of the "CO_2 anomaly." I had defined it as the difference between (1) the observed rise in CO_2 concentration during the last few thousand years and (2) the observed drops in CO_2 concentrations during previous interglaciations (fig. 11.1). Until this point, I had been focusing only on biomass burning that added CO_2 to the atmosphere, but now I realized that I had been ignoring the other side of the ledger: the natural drops in atmospheric CO_2 concentrations of 25 to 45 parts per million that had occurred during the four previous interglaciations. During the current interglaciation, a similar-looking drop had begun 10,500 years ago, but after 8,000 years ago the CO_2 trend had reversed direction and begun the anomalous rise. It occurred to me at this point that the *prevention of a natural CO_2 drop* should also count as a contribution to the size of the CO_2 anomaly. If part of the anomaly was explained by the lack of such a drop during recent millennia, then the part of the anomaly produced by direct carbon emissions from biomass burning would not have to be as large.

What were the mechanisms that had caused the natural CO_2 decreases during the earlier interglaciations but had been prevented from doing so during the present one? Two possibilities came to mind, both based on well-known hypotheses published in the early 1990s, and both consistent with the evidence from ice cores and marine sediments.

One potential mechanism is that advances of Antarctic sea ice cut off carbon exchanges between the ocean and the atmosphere and cause decreases in atmospheric

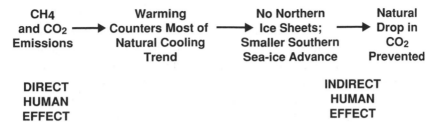

11.4. The CO_2 anomaly caused by humans prior to the industrial era was due in part to burning of forests (direct CO_2 emissions) and in part to the prevention of a natural CO_2 decrease.

CO_2 concentrations. The evidence cited earlier from Antarctic ice cores fits this explanation nicely: major decreases in Antarctic temperature occurred early in previous interglaciations, but only a small cooling has taken place during the current one. Sea ice would have advanced and CO_2 levels would have fallen during the previous interglaciations for this reason but would have failed to do so during this one.

The other mechanism is related to CO_2 changes caused by ice sheets in the North. Scientists have hypothesized that ice sheets create their own positive CO_2 feedback, forcing CO_2 levels lower as the ice grows, and letting concentrations rise as it melts. The means by which the ice sheets affect CO_2 levels is still debated, but two promising candidates have been identified. One hypothesis is that the dust generated on the continents by the ice sheets is blown to the ocean and fertilizes algae in the surface layers of the ocean. As the algae die, their carbon-rich tissue sinks to the sea floor, removing CO_2 from the surface ocean and the atmosphere. The other hypothesis calls on large ice sheets to alter the circulation and chemistry of the deep ocean in such a way as to change the CO_2 content of the atmosphere. Again the evidence cited earlier in this chapter is consistent with this explanation: ice sheets that grew during the early stages of previous interglaciations presumably helped to drive CO_2 levels down, but the failure of ice to appear during the current interglaciation should have kept CO_2 levels higher.

My proposed answer to the carbon-budget CO_2 dilemma is shown in figure 11.4. Emissions of carbon from human activities accounted for most of the observed methane anomaly, but only a fraction of the CO_2 anomaly (probably about a third). These "direct" emissions of methane and CO_2 caused climate to warm, and the warming suppressed a natural advance of sea ice in the South and prevented new ice sheets from beginning to grow in the North. A natural drop in CO_2 was thereby averted, and the absence of this decrease is the "indirect" part of the CO_2 anomaly. This explanation, if correct, resolves the impasse over the CO_2

budget. Because these indirect mechanisms are also entirely the result of human activities, humans are responsible for the CO_2 anomaly, as originally proposed.

In summary, my original hypothesis (the thesis) has already been adjusted in response to the challenges directed at it (the antithesis). The evidence that at first seemed inconsistent with the hypothesis turned out not to be so. In my opinion, the issues are now clearer and the hypothesis stronger because of these challenges.

As I write this chapter (a late addition to the book), I have just submitted a paper to a scientific journal summarizing all of these challenges and responses, and the community at large is still unaware of this work. It may take years for my arguments to persuade the scientific community. One reason for the gradual pace of acceptance is the reluctance on the part of scientists to abandon concepts that they have accepted and lived with for a long time. For decades, scientists have "known" several basic "truths." One is that climate has been warm for a little over 10,000 years, the length of the present interglaciation. Another is that this warmth has occurred for a natural reason: the forces that drive climate into new glaciations at regular cycles have not yet become strong enough to do so. A third "truth" is that human emissions of greenhouse gases became important only a few hundred years ago.

The first of these "truths" really is true. Climate has remained warm, cooling a little only in polar regions. But my hypothesis says that the second and third truths are not true. The warmth of the last several thousand years stems from a colossal coincidence: a natural cooling that was almost completely offset by a human-induced warming tied to greenhouse-gas emissions from agriculture.

A related reason that new ideas are resisted is simply time. At major research universities in the United States, scientists prepare and deliver lectures, and they counsel and advise undergraduate and graduate students. They serve on committees at departmental, university, national, and international levels. They write proposals for funding, and they manage and oversee labs, assistants, and students working on funded grants. They review proposals for funding agencies and papers for scientific journals. Week after week, two endeavors are shoved to the tail end of this long list of priorities: (1) reading and carefully scrutinizing new papers, especially in areas not relevant to their immediate research focus, and (2) thinking— the long, slow process of digesting and weighing enough information to allow new ideas to germinate. In a sense, the very structure of our work lives seems designed to keep us from achieving this most important goal of our research.

So it will take time for this hypothesis to be considered (by those who have the time) and for the evidence to overcome the inertia of "truths" scientists already "know." I feel confident that this acceptance will come.

Disease Enters the Picture

DISEASE ENTERS THE PICTURE

LOOK BACK AT the CO_2 trend during the last 10,000 years in figure 11.1B. One odd thing is that the rise in CO_2 values slows in the last 2,000 years compared to the preceding millennia. This seems strange, given that human populations had increased and technological improvements had continued over this interval. Why wouldn't CO_2 concentrations have risen even faster? Even odder was the fact that the CO_2 values began oscillating up and down, sometimes dropping by as much 10 parts per million from the general trend. Because these wiggles are much larger than any possible errors in measuring the CO_2 concentrations, they must be real.

One seemingly likely explanation was that these wiggles were the result of natural changes in climate occurring over decades to centuries. Short-term oscillations are well known from climatic records during times when Northern Hemisphere ice sheets were large, but also when ice sheets were absent from North America and Europe. The cause of these fluctuations is not completely understood, but two plausible factors are sporadic volcanic explosions and small changes in the brightness of the Sun. Other evidence, however, convinced me that these natural variations in climate could not explain the large dips in CO_2 during the last 2,000 years (chapter 12). It then occurred to me that the explanation might lie in some process tied to human activities. After all, the oscillations were superimposed on a rising CO_2 trend that I had concluded was caused by humans over thousands of years. Could the answer lie in a process that reversed that trend, perhaps one that had killed enough humans to reverse the process of gradual deforestation for several decades or centuries? This idea sent me into the history books, where I stumbled on the forbidding image of the four horsemen of the apocalypse. One of these (pestilence, or disease) proved to be the most plausible culprit (chapter 13).

Out of this exploration emerged one more challenge to current thinking: several times in the last 2,000 years, major pandemics had literally "put a plague on" (killed) tens of millions of people, reversing the gradual clearance of forests for agriculture, and contributing to short-term climatic cooling, including the Little Ice Age interval between 1300 and 1900. Disease had become a factor in climate change.

Chapter Twelve

BUT WHAT ABOUT THOSE CO$_2$ "WIGGLES"?

MOST OF THE ANTARCTIC ICE SHEET is a high polar desert receiving no more than an inch or two of snowfall each year, the polar equivalent of the arid core of the Sahara. Ice cores from such locations cannot capture the detail needed to reveal the short-term variations in CO$_2$ and other gas and solid constituents that occur over centuries or decades. But along the lower margins of the ice sheet, snowfall is heavier, and in favorably sheltered sites it is blown into thicker piles by strong winds. In these places, more detailed records can be recovered. Two such locations in Antarctica yielded the high-resolution records of the last 2,500 years shown in figure 12.1.

The shading shows a projection of the long-term increase in CO$_2$ over the last 8,000 years that I attribute to human activities. Not long after 2,000 years ago, that rising trend slowed and then was interrupted by several dips in CO$_2$. The two ice-core records in figure 12-1 disagree in places but indicate two or perhaps three CO$_2$ minima: a broad but not very deep one from about AD 200 to 600, a short one near 1300–1400, and a final minimum that is both deep and broad from 1500 until about 1750.

One reason the CO$_2$ measurements from the two ice cores disagree in the area of overlap is dating error. Of the two records shown, the one from Law Dome is superior: it contains layers of very fine ash particles deposited by volcanic explosions of known age. It was also deposited at a faster rate than the core from Taylor Dome. In any case, these records agree on one important point: negative CO$_2$ oscillations as large as 4 to 10 parts per million occurred within the last 2,000 years. Processes within the ice can smooth the true CO$_2$ record and reduce its amplitude, but it is difficult to create negative oscillations where none actually existed.

What could explain these drops? The most obvious explanation is that they are tied to natural changes in climate. In the last decade, one of the most exciting areas of climate research has been the discovery of abrupt oscillations that occur over decades, centuries, and millennia, intervals that are all much shorter than changes in Earth's orbit. For most of Earth's ice-age history, these short-term variations must have had a natural origin because they occurred well before any possible human effect on climate. When ice sheets were present, these oscillations tended to be very large: short-term temperature changes across Greenland and the North Atlantic amounted to as much as one-third of the difference between

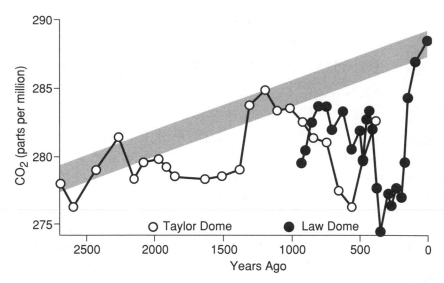

12.1. Antarctic ice cores show large CO_2 drops during the last 2,000 years compared to the long-term rising trend (shaded).

full-glacial and full-interglacial climates. During some of these oscillations, huge numbers of icebergs broke away from the coastal margins of the ice sheets, floated out into the Atlantic, and dropped enormous quantities of rock and mineral grains into the ocean sediments.

These relatively brief oscillations are superimposed on top of the slower, longer-term climatic changes at orbital cycles, and they appear to be the result of an entirely different phenomenon that is not yet well understood. Consider a simple analogy to something familiar. The daily (diurnal) heating cycle is driven directly by the Sun, with maximum warmth in late afternoon, and maximum cooling in the hours just before the Sun rises. These changes occur day after day in a predictable cycle. But on some days, especially in summer, storms build up in the afternoon, hide the Sun, drench the Earth with rain, and bring much cooler temperatures for an hour or two. In much the same way that these unpredictable storms briefly cool the predictable heat of the afternoon Sun, these shorter-term climatic oscillations over time scales of decades to millennia have ridden on the back of the longer and more predictable changes that occur over orbital time scales.

The reason these shorter oscillations are much smaller during times like today when the North American and Eurasian ice sheets are absent or reduced in size is not understood. The most recent large oscillation in climate occurred nearly 12,000 years ago, when the ice sheets were still large, followed by a smaller one

8,200 years ago, when the ice was almost gone. Since then, during the time humans created primitive and then advanced civilizations, these oscillations have been much smaller.

The largest interruption in this interval of relatively stable climate was the cooling during the Middle Ages called the Little Ice Age. This interval is variously considered to have lasted for as long as AD 1250–1900 to as short an interval as 1550–1850. The Little Ice Age is thought to have been preceded by a slightly warmer interval called the medieval climate optimum, with maximum warmth occurring around 900–1200. The Little Ice Age came to an end with the industrial-era warming of the 1900s that continues today.

In some places, the "Little" Ice Age was a very big deal indeed. Most such regions lie near the modern-day limits of Arctic snow and ice, where changes in climate are greatly amplified. For those alpine villagers who saw glaciers descend some 100 meters (more than 300 feet) down the side of mountains and grind their farms and villages to rubble, the Little Ice Age was certainly a big deal. These dramatic disruptions had been the basis for Louis Agassiz's realization that continent-sized ice sheets had once existed (chapter 4). Numerous etchings and some early photographs record the last of these events in the late 1800s. As the ice advanced, the upper tree line also descended. Intervening glacier retreats in prior centuries showed that the Little Ice Age was not a time of unrelieved cold, although on average it was cooler than today.

The coldest decades of the Little Ice Age were also a big deal for those farmers trying to grow frost-sensitive crops like corn and grapes at high latitudes and altitudes along the limits where such crops were barely possible even at the best of times. During some years and decades, the crops were destroyed by unseasonable freezes, or the harvest dates were delayed by summers with persistent cold or rain. Vineyards that had been started in Britain during the climatic optimum were progressively abandoned as the Little Ice Age intensified, and the northern limit for growing grapes in France and Germany retreated some 500 kilometers to the south. Cereals could no longer be cultivated in the hills of northern and western Britain.

Another place where the cold of Little Ice Age winters made a big difference was Iceland, a country that has long been highly dependent on its cod and other fisheries not just for commerce but also for basic food and survival. For many winter days in the Little Ice Age, the sea-ice pack surrounded the northern seaports and the fleet had to stay in port. Homebound Icelanders with little else to do kept reliable records of the number of days per year when the impassable barrier of ice blocked their fleets from reaching the sea (fig. 12.2). Sea ice was rare in AD 1000–1200, increasingly abundant between 1200 and 1800–1900, and rare again in the late 1900s.

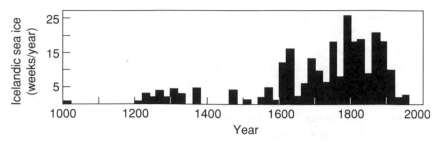

12.2. The number of weeks per year that sea ice jammed northern Iceland ports increased for several centuries but then dropped sharply during the 1900s.

But records of Icelandic sea ice and alpine glaciers speak for only a small part of Earth's surface, and local records may not represent the larger picture. For example, the winter of 1976–1977 was very cold by normal eastern United States standards, with harbors choked with ice south to New York and beyond. Two decades later, the winter of 1996 was the snowiest of the century for many eastern states. Yet these cold, snowy winters occurred during an interval when, on average, the planet has been gradually warming to unusually high levels. Even though the East Coast was having a few old-time winters (the kind your grandfather talked about), the planet as a whole was warming. Records from one or two regions do not necessarily reveal the larger picture.

As a result, some climate scientists expressed doubts about whether or not there really was such a thing as a Little Ice Age (or a medieval climatic optimum) on a global basis. This kind of challenge could be answered only by looking at detailed records of temperature changes from many regions. Unfortunately, high-resolution historical records that stretch back many centuries (like the one from Iceland) are extremely rare. So instead, climate scientists used techniques to extract climatic signals from natural archives.

The most common natural archive of climate change is the width of rings added each year by long-lived trees. In regions where summers are generally cold and somewhat inhospitable, trees add thicker rings during favorable conditions in slightly warmer (or wetter) years. Scientists spend summers in mosquito-infested areas of the Eurasian and American Arctic finding these long-lived trees and (non-destructively) extracting pencil-thin cores. Back in the lab, they count the rings to date each year of growth history and measure the ring thickness (or other properties) to extract records of yearly temperature change during the life span of the trees. Records like these now exist from far-northern latitudes all around the Arctic.

Other kinds of records of annual temperature change are available from the midlatitudes and tropics. At very high altitudes, annually deposited layers of snow harden into mountain-glacier ice that contains many kinds of climatic records at

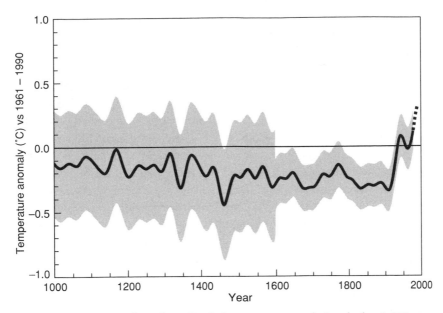

12.3. Reconstructions of Northern Hemisphere temperature during the last 1,000 years show an erratic cooling trend through the late 1800s followed by an abrupt warming during the 1900s. The light shading indicates uncertainties.

yearly or near-yearly resolution. Hardy scientists (like the legendary Lonnie Thompson) climb to altitudes of 6,000 meters (20,000 feet) or more, elevations near those familiar to professional mountain climbers, to drill cores into the ice. On the way in, they carry all their equipment and food. On the way out, they add hundreds of meters of newly drilled ice cores to their loads.

Not all of the work done to obtain annual records of climate involves such hardships. In the tropical oceans, some kinds of corals build their reef structures in annual layers that can be cored in warm, shallow, aquamarine waters. But most of the reefs that lie within easy reach of comfortable hotels have been cored by now, so that even this work requires travel in small planes to isolated islands that have few luxuries and in some cases host exotic tropical diseases.

From these natural archives, scientists have obtained climatic records many hundreds of years long at dozens of sites, mostly in the Northern Hemisphere. In 1999 atmospheric scientist Mike Mann and colleagues Ray Bradley and Malcolm Hughes used mathematical techniques to extract a record of estimated temperature change in the Northern Hemisphere over the last 1,000 years. The reconstructed trend (fig. 12.3) is known as the "hockey stick." The gradual cooling from AD 1000 through roughly 1900 is the handle of a hockey stick, and the much faster warming from 1900 through 2000 is its blade. Both the handle and the blade show significant shorter-term temperature oscillations, but the overall trend (at a quick squint) is much like a hockey stick. Many other reconstructions using records from different combinations of sites show the same basic trend.

At least in a broad sense, the reconstruction in figure 12.3 shows that the medieval interval around 1000 to 1200 was relatively warm compared to the gradual and somewhat erratic cooling in subsequent centuries. No one time stands out as the clear start of the Little Ice Age, but temperatures were certainly cooler by 1600–1900 than they had been earlier. If the warmer centuries of 1000–1200 are used as a baseline, the average temperature had cooled 0.1°C to 0.2°C by 1600 to 1900, and a few extreme decades showed changes as large as 0.4°C.

What could have caused this gradual, oscillating cooling trend after AD 1200? One factor might be the slow cooling of far-northern regions in response to orbital changes in solar radiation (chapter 10). Some parts of the Arctic have cooled by 1 to 2°C over 9,000 years. At that average rate, the cooling over 900 years would be 0.1 to 0.2°C, just about what the curve shows. But this cannot be the main explanation: the average drop in temperature across the entire Northern Hemisphere would have been much smaller than the amount at high latitudes (where responses are amplified), probably well under 0.1°C. Other factors must also be contributing to this cooling.

One likely factor is large volcanic explosions, which inject small amounts of sulfur dioxide more than 15 kilometers high into the atmosphere, well above the tops of clouds. The sulfur gas reacts with water vapor already present in the atmosphere and forms sulfuric-acid droplets that reflect some incoming solar radiation and keep it from reaching Earth's surface. As a result, the climate cools. Temperatures remain cooler for the two years or so it takes Earth's gravity to pull the tiny particles back into the lower atmosphere, where snows and rains can quickly wash them out. One eruption can cool climate noticeably for a year or two, but several eruptions bunched within a few years are required to keep climate cool for as long as a decade.

Tropical eruptions can cool the entire planet by spreading sulfur into both hemispheres, but eruptions to the north or the south of the tropics cool only the hemisphere in which they occur. Ben Franklin, taking note of the colorful sunsets and unusually cool summer temperatures in 1784 just after an eruption in Iceland, realized that volcanic explosions might have a cooling effect on climate. After the 1883 eruption of Tambora, crops failed in New England because of late and early freezes, and the locals referred to that year as "Eighteen Hundred and Froze to Death." Many of these earlier eruptions were much larger than the 1992 Mount Pinatubo eruption in the Philippines that gave us one summer of beautiful sunsets and cooled global climate by 0.3°C for a year.

The second factor proposed to affect climate over centuries or decades is the changing strength of the Sun (not to be confused with changes in the distribution of solar radiation caused by variations in Earth's orbit). Satellite measurements since 1981 have shown very small changes in solar radiation occurring with the same timing as changes in the number of dark spots (sunspots) on the Sun's surface. Although the dark spots reduce the amount of radiation streaming away from the Sun in the small areas where they occur, the overall relationship is just the reverse: the Sun emits more radiation when sunspots are most common. Stronger activity on other parts of its surface more than counteracts the localized effect of sunspots in reducing emissions.

Both the number of sunspots and the amount of solar radiation cycle from maximum to minimum size and back every 11 years. The 11-year cycle of sunspots is well recorded in telescope observations that go back into the 1700s, but almost no climatic records from Earth's surface show a convincing 11-year temperature cycle because the changes in radiation are so small. Yet this does not necessarily mean that changes in solar radiation are unimportant over longer intervals. Telescope observations from the Middle Ages show that clusters of 11-year sunspot cycles can vary greatly in amplitude, even disappearing altogether during intervals such as 1645 to 1715. Some (but not all) climate scientists believe that these intervals of less-frequent sunspots, and by inference weaker Sun strength, played a role in causing the cooler climates of the Little Ice Age.

What about greenhouse gases? What is their role in all this? Clearly the fast warming at the blade end of the hockey stick must at least partly be caused by greenhouse gases; both the rate and the amount of rise are unprecedented compared to the first 900 years of the record. Part 5 of this book will explore this recent warming trend. But what about the earlier record, and particularly the CO$_2$ wiggles in figure 12.1? The most recent CO$_2$ minimum (from 1500 to 1750) seems to match at least broadly the coolest intervals of the Little Ice Age, and higher CO$_2$ values tended to occur during the slightly warmer medieval era between 1000 and 1200.

The conventional explanation for this correlation of temperature and CO$_2$ is that changes in volcanic explosions and/or solar radiation are the "first cause" of both the Northern Hemisphere cooling and the CO$_2$ decreases. In this view, CO$_2$ values drop when Earth cools largely because of a basic law of chemistry: more CO$_2$ can be absorbed in a cool ocean than in a warm one. If volcanic and solar changes cool Earth's climate, the cooler ocean takes CO$_2$ from the atmosphere. If this view is correct, both the hemispheric temperature changes and the CO$_2$ variations are simply two responses of the climate system to natural (solar and volcanic) changes.

But this plausible-sounding explanation has a serious flaw. Today's most advanced climate models are constructed with the goal of reproducing all the complex interactions among the atmosphere, ocean, land surfaces, vegetation, snow, and ice. These models attempt to simulate all of these interconnected responses, rather than analyzing them one-by-one in isolation. When scientists specify changes in the amount of solar radiation entering the climate system as a result of volcanic eruptions and/or changes in Sun strength, the models simulate the integrated response of the many parts of the climate system, including the changes in temperature and atmospheric CO_2 levels.

Examining one of these modeling studies, I noticed a result that seemed to indicate a major flaw in the natural explanation. To match the largest (10 parts per million) decreases in CO_2, the model had to cool climate by almost 1°C, yet the reconstructed temperature trend in figure 12.3 permitted a decrease of only 0.1–0.2°C between the warmer interval from 1000 to 1200 and the cooler one from 1500 to1750. Conversely, to stay within the bounds of the small temperature changes in this reconstruction, the models would allow CO_2 changes of only 2–3 parts per million, compared to the 10 parts per million changes observed (fig. 12.1). Whichever way I looked at it, the CO_2 changes were simply too large relative to the temperature changes. Something seemed to be seriously wrong with the conventional explanation.

The natural explanation also struck me as suspect for another reason. The rates of change during these CO_2 drops looked very abrupt, in fact much faster than the natural changes that had occurred at the end of the last deglaciation, and during the previous three deglaciations as well. As shown in figure 9.1, CO_2 concentrations during each of these deglaciations rose by almost 100 parts per million within a span of not much more than 5,000 years, for an average rate of about 2 parts per million per century. In contrast, the CO_2 concentrations in figure 12.1 fell and then rebounded by 10 parts per million within only a century or so, a rate approximately five times as fast as the deglacial changes. Why would CO_2 values change at a faster rate during a time of almost stable global climate than they had during the highly dynamic climatic changes that occurred at the end of the major glacial cycles? This didn't make any sense.

So once again, an explanation based on natural climatic changes had come up short. And again it seemed to me that the only solution to this apparent dead end must be an explanation lying outside the realm of "natural" causes, an explanation somehow linked to humans. The explanation had to lie in some kind of process that could reverse the slow deforestation and accompanying CO_2 releases that had been occurring for thousands of years and could cause abrupt CO_2 decreases that lasted for decades to a century or two.

Chapter Thirteen

THE HORSEMEN OF THE APOCALYPSE: WHICH ONE?

HISTORIANS HAVE LONG SENSED that the last centuries of the Roman era, and those that followed, were something of a reversal in the "onward and upward" march of human progress typical of previous centuries, at least in Europe. The Romans had for a while achieved a level of engineering technology and general prosperity that would not be repeated again in most of the West for over a millennium. Aqueducts brought to their cities fresh water of a quality not equaled until less than 200 years ago in London or Paris. The aqueducts, as well as baths and public structures not intentionally destroyed by human hands, have stayed intact for almost two millennia because they were bound by cement of a quality that Europeans still could not match in the early 1800s. The durability of Roman roads was also unsurpassed. In most respects, life in Rome during the height of the Roman Empire had much more in common with life during the early 1800s than it did with conditions 8,000, 6,000, or 4,000 years ago. Similar advances had also occurred by 2,000 years ago in the advanced civilizations of China and India.

But soon after the height of the Roman era, Western civilization entered the "Dark Ages," a time marked in most areas by a regression in technological knowledge and a general loss of respect for, or interest in, scientific inquiry. In much of the West, this stagnant condition was to last until the Renaissance almost a millennium later. The Dark Age was also a difficult time in a more fundamental sense: world population growth slowed and even stopped during some intervals, apparently because of rising mortality rates. A woodcut in 1528 by Albrecht Dürer left an enduring image of several kinds of devastation that added to normal mortality during this era: the four horseman of the apocalypse. This image came originally from Revelations, the last book of the New Testament. The identities of the horsemen have been variously interpreted through history, but the typical version names three of them war, famine, and pestilence, with death the fourth.

The horsemen of the apocalypse seemed like a useful starting point to search for the cause of the unexplained CO_2 drops during the Dark Ages. Whichever horseman had killed the most humans might explain the CO_2 reductions: fewer humans, and less CO_2 release. Although not a historian, I began to explore the last 2,000 years of recorded human history, optimistically searching for any link to those dips in the CO_2 concentrations (see fig. 12.1), with the horsemen of the apocalypse as my gloomy guides.

An ancient, and somewhat fatalistic, wisdom acknowledges that human progress is not always a one-way process, and a popular song lyric from recent decades captures this attitude well: "one step forward, and two steps back." The initial discovery of agriculture, and all of the innovations that followed, had for several millennia been of unprecedented benefit to humanity. With more food available, human populations grew as never before. With dependable sources of food in a single region, people could stay in one place, rather than constantly having to pack up and move on. With their livestock nearby, people had easy access to protein-rich meat, milk, butter, and cheese to supplement crops rich in carbohydrates. With orchards of fruit and nut trees, diets became more nutritious still. With crop excesses stored away for lean years, extreme swings in weather could often be ridden out. Compared to the more vulnerable existence of those who had lived by hunting, fishing, and gathering, agriculture had transformed existence on Earth.

But agricultural progress came at a cost. Embedded in this new way of life were changes that would lead to problems on a scale previously unknown in human history. Each of the first three horsemen of the apocalypse is, indirectly or directly, linked to the unprecedented success of agriculture. And gradually, after 2,000 years ago, the "forward" steps of agricultural success would begin to produce "backward" feedback effects that would grievously afflict humanity.

Consider war, obviously a major killer of humans. By 2,000 years ago, war was occurring on previously unprecedented scales. Stone Age clans moving from site to site in forest clearings no doubt fought often over resources, but they did so locally, with deaths on a smaller and more random scale. But after agriculture produced far larger societies with much greater wealth, these cultures began to pay a class of full-time warriors to defend that wealth and to invade other regions to obtain even more. In addition, as different religions came into being from region to region, differences in beliefs became a common motivation for war.

If war had been the major cause of human mortality, I would have expected to see the largest wars clustered within intervals of low CO_2 concentrations at roughly AD 200–600, 1300–1400, and 1500–1750, with little warfare from 600 to 1300. Yet a cursory look at the history of warfare showed little or no obvious correlation of this kind. We (humans) have rarely allowed much time to pass without a major war, and no era of any length has been entirely free from it on one scale or another. An animated world map of the history of warfare would show endless overlapping battles of considerable intensity throughout every interval of history. To my eye, no convincing link to the three intervals of low CO_2 was apparent.

Some might still be tempted to infer that a significant concentration of wars occurred during the decline of the Roman Empire from 200 to 600, the first low-CO_2

interval. The Huns, the Visigoths, and then the Vandals invaded southward from Germany beginning in 370 and through the middle 500s, followed by the Avars in the 500s to 600s and the Slavs in the 500s to 700s. Large disturbances also occurred in China and India during this time. But war did not subsequently cease for the next 600 years when CO_2 levels rose. Later warring groups included the Franks in the 700s, the Vikings in the late 700s to the 900s, the Magyars in the 800s and 900s, and the Muslims from the 600s and until the Saracen resurgence of the 1200s. With no obvious gap in the frequency of warfare, little correlation to CO_2 emissions was apparent.

One important exception may exist: the Mongol invasions throughout Eurasia that began in the early 1200s and reached a peak during the middle 1200s left effects that lasted for centuries afterward. Genghis Khan and his successors invaded every region from China in the East, to India in the South, to Europe in the West, and in several regions they ripped up the very fabric of the societies they conquered. Tens of millions may have died in China at Mongol hands in the 1200s. In the arid Near East, where agriculture had originated, the Mongols destroyed most of the existing irrigation-based agriculture, and populations fell precipitously. War and systematic destruction at this large a scale is unusual, and CO_2 concentrations were falling during this interval, so a causal link is possible in this case.

But in all other respects, war seemed an unlikely explanation of the CO_2 drops during preindustrial times, at least to my nonhistorian's eye. Even the 8 million Germans and Belgians who died during the Thirty Years War between 1618 and 1648 were "only" 1–2% of the global population of that time. War is deadly, but seemingly not deadly enough to qualify as the horseman I sought.

Famine is another potential killer, and to some extent also an outgrowth of the success of agriculture. With gradual improvements in agricultural techniques, farmers gradually began taking the risk of growing crops closer to the limits nature sets: in far-northern regions and on mountain flanks where cold temperatures set natural limits, and in warmer semi-arid regions where drought is the limit. Crops grown in these environments naturally became more vulnerable to year-to-year freezes or droughts, and even more so to longer-term climatic changes. The most severe famine in preindustrial European history in terms of mortality occurred in 1315–1322, and another in the 1430s.

But was famine really a major factor on a global scale? Like today, very small fractions of the total human population on Earth lived along the northern or high-altitude limits of agricultural regions. When crops failed in these cooler regions, the deaths they caused tended to be relatively small on a global scale. Even the famine of 1315–1322 mainly affected far-northern and high-elevation regions of Europe. By 1322 good crops were again coming in, and population levels seem to have quickly recovered.

What about the tropics and subtropics, where most humans actually live and where the most serious climatic concern is drought? Could drought-induced famine be a major killer across large areas of the tropics and subtropics? This possibility seems unlikely for several reasons. For one thing, irrigation provided much of southern Eurasia with a buffer against the worst impacts of drought. Many of Eurasia's agricultural regions had dependable supplies of water from rivers that flowed from well-fed mountain sources that received considerable rainfall even during droughts.

In addition, the likelihood of drought striking vast areas of Eurasia simultaneously is unlikely on a meteorological basis. On a global scale, very nearly the same amount of rain falls each year. Solar radiation evaporates very nearly the same amount of water vapor from the oceans and land each year, but the atmosphere can store only so much water vapor before it sends it back to Earth. As a result, global rainfall does not vary much from year to year.

Of course precipitation does vary widely on a local basis. One town can be as green as a golf course late in June, with another town nearby parched and brown; yet a month later the two regions may have switched colors, as the scattered thunderstorms that first favored one area then fell on the other. This same uneven distribution of rainfall occurs at larger regional scales, with one country suffering a multiyear drought while a neighbor has heavy rains and flooding. But on a larger, more nearly global scale, the droughts and floods (and normal rainfall elsewhere) tend to balance out. It is nearly impossible for a region as large as the entire southern tier of Eurasia to be gripped in drought at a single time. To my knowledge, no climate historian has ever claimed that drought has simultaneously afflicted this largest and most populated of continents during the last 2,000 years. Because neither freezes nor droughts are likely to produce simultaneous crop losses across large portions of Eurasia, famine does not seem to be the horseman I sought.

What about the horseman called pestilence, or disease? One afternoon when I had just begun wondering about those CO_2 dips, I was eating lunch and reading a book review when the word "plague" caught my eye. I put down my sandwich and walked over and pulled out the encyclopedia. There, I quickly relearned that bubonic plague had caused the Black Death pandemic of the mid-1300s, as well as later outbreaks during the 1500s and 1600s, but I also learned for the first time that a major pandemic had occurred during the Roman era in AD 540–542. The rates of mortality in both pandemics had been incredibly high (killing over 25% of the population), and at first glance the pandemics correlated fairly well with the dips in CO_2. Here was a more promising explanation of the CO_2 drops: pestilence, the rider of the pale horse.

As Jared Diamond summarized in *Guns, Germs, and Steel*, the very successes of agriculture had also been favorable to the spread of disease. In earlier times, people living the hunting-fishing-gathering life were dispersed in small clans or tribes. If disease struck a clan or local group, some (or even most) of its members might die, but the likelihood of them transmitting it to other clans was limited. Hearing about strange deaths in one group, people nearby could flee to areas beyond reach of the disease. People died from disease, but primarily at the limited scale of clans or tribes.

By 2,000 years ago, the rapid growth of populations had eliminated some of this natural protection. Ample production of food in the populated regions of eastern Asia, India, and Europe had led to the growth of towns and then cities. Because dense concentrations of people are natural breeding grounds through which contagious diseases can be transmitted quickly, both the victims and the carriers of virulent diseases were now conveniently clustered close together.

In addition, the fact that farmers lived settled, sedentary lives helped to breed disease. In preagricultural times, when food sources were depleted, hunting-gathering clans had been forced to move to new sources, leaving their refuse behind. Now, with large food surpluses from agriculture, people could live in one place and in ever-greater numbers, and their dwellings were surrounded by growing amounts of rubbish and waste. Permanent houses attracted mice and rats, carriers of diseases, and in many towns and cities the sanitation was primitive and the streets strewn with refuse, excellent breeding grounds for disease. Human feces were spread on fields for manure, and even irrigation ditches became potential sources of contagion.

Even worse, the livestock that humans had begun tending thousands of years earlier were carriers of diseases that afflicted people. Cattle carry smallpox, measles, and tuberculosis. Pigs carry influenza. Another factor indirectly related to agriculture was the increase in travel because of improved ships and greater use of overland routes for trading goods. Agricultural success generated increased wealth and increased trade, and trade put regions in closer contact. Now, when disease hit one area, it could more easily be carried to others. For all of these reasons, agricultural successes gave disease wider access to human victims.

Until the historical era, little is known about disease. The Old Testament mentions pestilence several times (for example, in First Samuel). As the historical era began, written records of disease in some regions give us a limited picture of this history. I had initially hoped to find a graphic plot of disease mortality through the historical era, but I have not yet succeeded, although such a plot may well exist somewhere. Instead I compiled my own plot based mainly on *Plagues and Peoples* by W. McNeil, *Disease and History* by F. E. Cartwright, and *Armies of Pestilence* by R. S. Bray.

An accurate, reliable plot of the entire history of disease mortality seemed impossible. Instead, I chose to portray a general sense of the scope of mortality

TABLE 13.1
Epidemics and Pandemics of the Last 2000 Years

Year (AD)	Region	Disease	Intensity (% Mortality)
79, 125	Rome	Malaria?	Local epidemic
160–189	Roman Empire	Smallpox?	Regional epidemic
265–313	China	Smallpox	Regional epidemic
251–539	Roman Empire	Smallpox? or bubonic plague?	Regional epidemics (decadal repetition)
540–590	Europe, Arabia, and North Africa	Bubonic plague	*Major pandemic* (25%) Decadal repetition (40%)
581	India	Smallpox?	Regional epidemic
627–717	Middle East	Bubonic plague	Local epidemics
664	Europe	Bubonic plague	Regional epidemic
680	Med. Europe	Bubonic plague	Regional epidemic
746–748	Eastern Med.	Bubonic plague	Local epidemic
980	India	Smallpox	Regional epidemic
1257–1259	Europe	Unknown	Regional epidemic
1345–1400	Europe	Bubonic plague	*Major pandemic* (40%)
1400–1720	Europe/North Africa	Decadal repetition	Regional epidemic
1500–1800	Europe	Smallpox	Regional epidemic
1500–1800	Americas		*Major pandemic* (80–90%)
1489–1850	Europe	Typhus	Regional epidemic
1503–1817	India	Cholera	Local epidemic
1817–1902	India/China/Europe		Pandemic (< 5%)
1323–1889	Europe	Influenza	Regional epidemic
1918–1919	Global		Pandemic (2–3%)
1894–1920	Southeast Asia	Bubonic plague	Regional epidemic (small %)

through time by plotting historically recorded outbreaks on three spatial scales: local epidemics, which affected towns or parts of countries; regional epidemics, which affected several countries or small parts of more than one continent; and pandemics, which affected large parts of several continents. The major disease outbreaks are listed in table 13.1 and plotted alongside changes in ice-core CO_2 and the populations of disease-afflicted continents in figure 13.1. The population

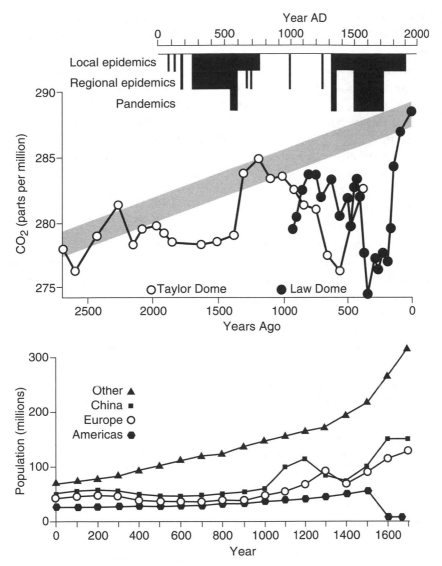

13.1. Intervals of low CO_2 concentrations in Antarctic ice cores correlate (within dating uncertainties) with major pandemics that decimated populations in Eurasia and the Americas.

numbers for Eurasia come from McEvedy and Jones's *Atlas of World Population History*. In my opinion, the correlation between pandemics, population losses, and dips in CO_2 minima looked suggestive enough to be worth pursuing.

The historical record of disease began when a pestilence of unknown origin struck Athens in 430 BC, decimating the Athenian army during the Peloponnesian

War with Sparta. Outbreaks of what may have been malaria occurred in Italy in AD 79 and 125 (the latter called the plague of Orosius). Malaria is not usually fatal today, but these outbreaks killed large numbers of people, perhaps because natural resistance to it had not yet developed. The disease hit especially hard in the countryside, and many farms went out of cultivation as farmers moved into the cities. Still, these outbreaks seem to have been relatively local in scale. A somewhat more extended outbreak, apparently of smallpox, the Antonine Plague, struck the Roman Empire between 160 and 189, killing the Roman emperor Marcus Aurelius. At its peak, some 2,000 people died each day. The symptoms (fever, skin eruptions, and inflammation of the mouth and throat) match those of smallpox. Mortality rates were high enough that farmers were drafted to fill the depleted ranks of the Roman army.

The year 251 saw a renewed pestilence called the plague of Cyprian after one of its early victims. The type of disease is again uncertain, but smallpox and bubonic plague are among the more likely choices. Whatever its origin, it was lethal and lasted for 16 years, becoming nearly pandemic from Egypt to Scotland. Anecdotal information from some regions indicates more deaths than survivors in the wake of this pestilence, with large areas of farmland in Italy reverting to the wild. Once again, people fled from the countryside to the cities, where they may have been more vulnerable. Recurrent outbreaks of bubonic plague continued in the Roman Empire for three centuries, and some (but by no means all) historians infer that the Roman Empire began to be seriously weakened by this ongoing loss of population.

Some historians also claim that the extended sequence of pestilence between the first and fourth centuries AD contributed to the growth of Christianity. As many people grew ill and died, the terrified survivors would have found little consolation in the jealous infighting of the older Roman gods. By contrast, Christianity offered the hope of miracles in this life (healing of the sick and casting out of spirits) and the promise of life after death. As conversions to Christianity accelerated during the plagues, it moved from an outlawed religion of martyrs to the official religion of the empire by the end of the fourth century.

But the pestilence continued during Christian times, culminating in 540–542 with a plague of unprecedented intensity named after Justinian, emperor at that time and destined to be the last of the line. This pandemic, almost certainly bubonic plague, was first recorded in Egypt, swept north into Palestine, Greece, and then the Black Sea and Constantinople, by that time the center of authority for the remnants of the empire. It then moved through the cities of North Africa along the Mediterranean coast and into southern and western Europe. It generally arrived first at coastal seaports and then spread into the interior cities and countryside. Because its lethal effects were felt on several continents, this plague was the first great pandemic.

Most of the plague during this era was probably carried by fleas hitching rides on rats. The fleas bit humans and transmitted the disease, which began with a fever, followed by large swellings (buboes) in the groin or armpits, and then a coma. Any physical movement on the part of those afflicted caused excruciating pain. Shortly after death, black spots appeared on the body. In one sense, at least, bubonic plague had a merciful side: death came quickly, usually within a week. At the height of the plague of Justinian, 5,000 to 10,000 people died each day in Constantinople, too many to be buried by the survivors. Their rotting bodies, filling the cities with the stench of death, were dumped into the empty towers of forts or loaded onto ships that were set adrift on the ocean. This time, not just villages and towns but even some cities ceased to exist, as the social order largely collapsed, and agriculture almost entirely ceased in many regions. Some 25% of the population may have died in this one outbreak.

Lesser but still extremely severe outbreaks of bubonic plague continued in Europe at intervals of 10 to 20 years until AD 590. The reason for the roughly decade-long spacing of outbreaks may be that most survivors had a natural resistance to the disease, and so the plague briefly abated for lack of available victims. After 15 or 20 years passed, many members of the newest generation lacked immunity, and plague would flare up anew. Some 40 percent of the population of Mediterranean Europe may have died in the cumulative outbreaks by 590 (fig. 13.1). For some countries, these losses were not replaced for four or five centuries.

For unknown reasons, the disease began to abate in Europe after 590, with regional epidemics recorded for the next 150 years, even though nearly continuous outbreaks of plague continued to afflict the Muslim-dominated Middle East between 627 and 717. After an outbreak in 746–748, bubonic plague disappeared for some 600 years.

The long Roman-era interval of epidemics that culminated in a major pandemic matches fairly closely the first extended CO_2 minimum in the ice-core record (fig. 13.1). The subsequent plague-free interval from 749 to the middle 1300s also correlates reasonably well with the rebound of the CO_2 trend (given the dating uncertainties in the latter), and of European and Chinese populations, to higher values. At least at a glance, it seems plausible that pandemics, populations, and CO_2 levels may have been linked during this interval.

Another devastating plague pandemic struck in the late 1340s. Originating perhaps in Central Asia, it reached the Near East by 1347 and swept across Europe from the Black Sea to the British Isles and into North Africa by the early 1350s. This plague was transmitted not just by fleas and rats, but also by pneumonic bacilli spread by coughing, sneezing, kissing, and even just breathing. Of an estimated 75 million people living in Europe just before the plague hit, at least 25 million, or one out of three, died within a few years.

Once again, farms, small villages, and even entire towns were abandoned, their inhabitants having died or fled to the cities. Crops again lay unharvested in the fields, and vineyards untended. As before, the cities provided no refuge, with mortality rates as high as 70 percent in the hardest-hit areas, although some regions were spared. Bodies of people from lower-class families were dragged out into the streets and left to rot, and few dared come to funerals held for the wealthy. The Pope was forced to consecrate the Rhone River near the city of Avignon so that bodies dumped there could be said to have received a Christian burial. Ships that had lost their entire crews to plague drifted aimlessly across the Mediterranean and North seas.

Some historians believe that the Black Death changed the feudal structure of medieval England, and perhaps elsewhere. Prior to the plague, poor serfs worked the land of their lords without much hope of improvement in their position. But after plague killed so many people, the resulting shortage of farm workers gave the survivors some bargaining power. For the first time, laborers moved around the countryside looking for higher wages and better situations. A form of tenant farming took hold—not complete freedom, but better than serfdom.

This first plague pandemic was followed by several more virulent outbreaks through the 1390s. By then, an estimated 40–45 percent of the population may have been killed in many parts of Europe. Entire villages simply disappeared, some forever. The impact of these plagues lingers today in phrases we use without giving their origin any thought: "avoiding someone like the plague"; a problem that "plagues us"; or "wishing a plague upon someone."

I find it difficult to imagine the horror of a disease that suddenly arrives from unknown sources by unknown means, kills an average of one out of three of your family members and your neighbors in a year or two, and then abates, at the point when you had begun to give up hope of surviving. Even more cruelly, by the time you have finally begun to feel safe and perhaps guardedly hopeful, the disease returns 15 or 20 years later and claims still another generation of victims. The Black Death pandemic and subsequent plague outbreaks through 1400 line up well with a dramatic CO_2 decrease in the Taylor Dome ice-core record, although the better-dated ice-core record from Law Dome shows a much less obvious drop (fig. 13.1).

Plague outbreaks continued in Europe for the next 300 years. Some historians refer to the worst of these outbreaks as pandemics, and thousands of people died in London each week at the height of an outbreak in 1665, just one of many during this interval. And by this point in human history, other diseases had also joined the "army of pestilence" in taking a toll on human populations. Influenza (from pigs) became a major problem by the early 1300s. Smallpox (from cattle) became epidemic in Europe in the 1500s and remained a major killer until the

1800s. Regional epidemics of cholera struck India after 1500, with an unusually lethal outbreak in 1543.

Yet this disease-ridden interval in Europe does not seem to rank as a true pandemic. After the Black Death horror of the mid-1300s, European populations rebounded to preplague levels within a remarkable 150 years and continued to grow from the 1400s through the 1800s (fig. 13.1). Mortality rates must have been considerably smaller than during the Roman or medieval eras.

But a third (and worst) preindustrial pandemic was still to come—the one resulting from European entry into the Americas. Europeans carried many diseases to which they had gained a large measure of immunity, but which decimated Native American populations from Canada to Argentina. Some of the diseases were carried on the persons of the Europeans, at that time mostly a flea-infested, lice-ridden people who abhorred bathing as unhealthy. Diseases also arrived on their pigs, cattle, and other livestock. This wave of invading pestilence, unprecedented in history, included smallpox, influenza, viral hepatitis, diphtheria, measles, mumps, typhus, and whooping cough, and somewhat later, scarlet fever, cholera, and bubonic plague. Even diseases such as mumps and measles that sound trivial today were often fatal for people with no natural immunity.

In recent decades, estimates of the size of the pre-Columbian populations in the Americas have soared. Where once historians thought that 10 to 20 million people lived in the Americas, conservative estimates from reputable sources like W. Denevan (*The Native Population of the Americas in 1492*) are now in the range of 50 to 60 million people, with more extreme proposals exceeding 100 million. The largest populations were the Aztecs in Mexico, the Inca in Peru and Bolivia, and the surprisingly large populations living in the tropical rain forests of Central America and the Amazon Basin. The much larger population estimates for the Amazon in recent years come from new archeological methods such as low-level airplane overflights that trace out road and village patterns or terraced hillside gardens, followed by detailed studies on the ground.

After the Europeans made contact, up to 90 percent of the native populations died. Entire villages that once lined the valleys of the lower Mississippi River system were abandoned, along with endless cornfields in between. After the forests again took over, the only obvious evidence left of the former existence of these agricultural people was massive earthen mounds used for ceremonial purposes. Most of these mounds were plowed by settlers and flattened to create towns and cities. In the Amazon Basin and other rain forest regions, lush tropical vegetation swallowed up most evidence of former habitation. Many decades later, so little evidence remained of the former occupation of North America that scientists and historians in the 1800s and early 1900s assumed that populations had been relatively small. For the few regions (like Cahokia, Illinois) where massive structures indicative of more advanced civilizations survived, scientists and historians discounted the obvious explanation (that they were of Native American origin) and assumed that European people must have created these structures in an earlier era.

Today, the best estimate is that some 50 million people died just from having come into contact with Europeans. This was the greatest pandemic in all of preindustrial history, and, in proportion to the size of the global population, the worst pandemic of all time. Out of roughly 500 million humans then alive on Earth, 50 million (10%) died in the Americas.

The indigenous populations of the Americas never recovered from this great pandemic. Not until large-scale European settlement after 1750 did the total population in the Americas reach the pre-Columbian level. The overall duration of this third great "American Pandemic"—1500–1750—closely matches the third and largest CO_2 drop in the Law Dome ice core (fig. 13.1). In this case, the age of the CO_2 minimum in the ice core is firm, and its correlation in time with the pandemic is certain.

Since the 1800s, mortality from diseases has been lower than before, with some exceptions. The last recorded epidemic of bubonic plague occurred in southern France and northern Africa in 1720. The near-disappearance of plague during the 1700s is often credited to improved sanitation, but this explanation is disputed by some medical historians who infer that the species of rat that carried plague-bearing fleas was for some reason displaced by a species that did not. Vaccines for bubonic plague became available only in 1884, but a serious outbreak still occurred in 1910 in Manchuria, and isolated cases occur even now. Gradually, improvements in sanitation and new medicines have suppressed the worst forms of many diseases, although AIDS has killed millions of people despite the best modern medical efforts. In any case, the rebound from low CO_2 levels after 1750–1800 seems to correlate reasonably well with the reduced incidence of high-mortality disease.

Historians must weigh and balance many contending explanations for the complex array of developments that determine the course of history. Still, the evidence in hand leads me to a clear conclusion: the major CO_2 dips in the ice-core records correlate more persuasively with population drops caused by major pandemics than they do with times of war or famine.

It was tempting to conclude that pandemics must be the primary *cause* of the CO_2 drops. But an old adage in science holds that "correlation does not prove causality." Two trends may be wonderfully correlated in time and yet not related in a cause-and-effect sense. It was still possible that the CO_2 values and the disease/population trends were each responding to some other common factor but were not actually linked in a causal sense at all. I still needed to find a specific, plausible causal mechanism that linked CO_2 and pandemics.

Chapter Fourteen

PANDEMICS, CO_2, AND CLIMATE

THE CORRELATION BETWEEN pandemics and drops in atmospheric CO_2 concentrations was suggestive, but what was the connection? How could plague and other diseases cause the drops in CO_2? Part of the answer to these questions comes from historical records summarized in chapter 13. These records document abandonment of farms and farm villages on a massive scale during and after all three major pandemics. In the wake of the European plagues, abandoned farms are described as having gone to waste or ruin. Those words bring to mind doors flapping in the breeze, roofs sagging and collapsing in upon houses and barns, and wild vines creeping up and strangling rotting fences. But nature was doing much more than that. Nature was busy turning pastures and croplands back into forest, and remarkably quickly.

A little over a decade ago, we bought the property where we now live. The former owner had begun a small Christmas tree farm in the lower part of the meadow where our house stands. He kept most of the meadow in good trim by "bush hogging," a method far cruder than cutting hay. The drive from a tractor spins a blunt blade that whacks down the grass by brute force, along with anything else that sprouts (saplings, shrubs, etc). Two summers had passed between the time when we bought the land and the point when house construction started, and in that short a time the meadow had begun to be invaded by forest. Young cedar trees sprouted along the meadow edges where berries fell from mature trees in nearby woods. Locust saplings sprouted well out into the meadow by a combination of root propagation and seed dispersal by pods. Several kinds of shrubs and cedars that had been whacked down but not killed by bush hogging also started to grow here and there. This meadow had no fences, or otherwise the birds would have started rows of trees on the ground just below by depositing seeds in ready-made fertilizer capsules. To me, a product of the suburbs, this instant eruption of trees was astonishing. It made me realize how much work it takes to keep a meadow a meadow. You can always put livestock out to browse down the invading vegetation, but only goats eat everything in sight. Extra cutting is needed to fight off tree and shrub invasions in fields with only cattle or horses.

Here in Virginia, cedars and locusts are the meadow invaders. To the north, in New York, maples are the aggressors. In northern New England, it's the birches. Looking at the remains of the abandoned farms throughout New England,

Robert Frost might just as well have written "Something there is that doesn't love a meadow." In all naturally forested regions, forests can reoccupy abandoned pasture or cultivated land with amazing speed.

History shows that many farms in Europe were not reoccupied for many decades or even a century or two after major plagues. The amount of time until reoccupation depended in part on whether or not the plague outbreaks repeated, thereby keeping populations low and people off the available land. In the absence of humans, vegetation began to reclaim the farms, and as it grew it pulled CO_2 out of the atmosphere. Based on field evidence and models, ecologists have found that abandoned pasture and cropland regain the carbon (biomass) levels typical of full forests within 50 years. At first, the invaders are shrubs and tree saplings competing for sunlight and soil. In time, some trees begin to out-compete others, and the thickets thin out a bit. After 50 years, sizeable trees exist, not old-stand forests with mature trees, but full forests nevertheless in terms of the amount of carbon stored in roots, trunks, and branches. From the perspective of the CO_2 being removed from the atmosphere, it is as if the land has completely reverted to forest in just 50 years.

Here, then, is a possible mechanism to pull CO_2 out of the atmosphere in a few decades: widespread reforestation as a result of pandemic mortality. Imagine the following scenario. During the years before the plague hits, gradual deforestation has been slowly removing carbon from the land at rates typical of the interval between 8,000 years ago and the industrial era. As a result, atmospheric CO_2 levels have been slowly rising. Then a pandemic strikes, causing mass human mortality, widespread farm abandonment, forest regrowth, and CO_2 removal from the atmosphere over the next 50 years. The rates of CO_2 removal in reforestation are much higher than the very gradual rates of CO_2 addition from deforestation, and 50 years or so after the onset of the plague, CO_2 concentrations have fallen to a minimum value.

At this point, several scenarios are possible. If the pandemics end, people will soon begin to reoccupy the farmland, cut back the newly grown forest, and restore the farmland to agricultural use. As they do so, carbon will be returned to the atmosphere, and CO_2 levels will rise rapidly back to the preplague trend. But if repeated outbreaks of plague occur at decadal intervals following the initial outbreak, over a century could pass before the farms begin to be reoccupied.

History shows that the patterns of the major pandemics were all different. The plague of Justinian in AD 540–542 was the culmination of several centuries of prior epidemics of increasing intensity, and it was followed by more than a century of less-severe outbreaks. Populations in most of Europe didn't fully recover until the start of the medieval era around AD 1000, many centuries later. In contrast, the Black Death struck without warning in 1347–1352 and was followed by several outbreaks, yet populations in most of Europe had somehow rebounded to previous levels by 1500. The American Pandemic after 1500 had no real end: indigenous populations had not even begun to recover when major European settlement started in the middle and late 1700s.

One aspect of the Roman-era and Black Death pandemics that struck me as curious remains unexplained. During the same centuries that plague was decimating Europe, populations in China were falling and then recovering in almost identical patterns (see fig. 13.1). Because China and much of Europe had numerous census counts throughout these intervals, this striking similarity in population changes is hard to reject as some kind of artifact of historical inaccuracies. Yet most of the primary sources on the history of disease do not mention bubonic plague in China during these intervals, or anywhere else east of Iran.

In any case, I omitted China from my assessment of the effects of reforestation on atmospheric CO$_2$ during both the Roman and medieval eras. Largely anecdotal historical evidence indicated that all of the arable land in China had been deforested some 3,000 years ago, at least in the populous north-central regions. By that time, China had passed the population density of 11 people/km^2 that caused near-total deforestation of England in 1089 (the Domesday Survey). I reasoned that the large excess of people available in China would have occupied any farmland abandoned during the population drops of the Roman and medieval eras, so that reforestation would have been minimal despite the enormous mortality at those times.

This left Europe and other circum-Mediterranean lands as the regions of likely reforestation during the first two pandemics, and the Americas during the third. To estimate the amount of farmland abandoned, I again turned to the concept of a per capita footprint of each human on the forest (chapter 9) but this time used it in the opposite sense—to quantify farm abandonment. Since each Stone Age or Iron Age person occupied an estimated 0.03–0.09 km^2 of land, I assumed that each person killed by plague would also leave that much land abandoned. Regrowth of forests on those farms would, after 50 years, pull as much carbon out of the atmosphere as the earlier forest clearance had once put into it. The total carbon pulled from the atmosphere is calculated as:

$$\begin{pmatrix} \text{\# people killed} \\ \text{by disease} \end{pmatrix} \times \begin{pmatrix} \text{farmland abandoned} \\ \text{per person} \end{pmatrix} \times \begin{pmatrix} \text{C per km}^2 \\ \text{reforested} \end{pmatrix}$$

The carbon sequestered in the growing trees does not come entirely from the atmosphere. The atmosphere constantly exchanges carbon with the surface layer of the ocean and with all of the world's plant matter (both terrestrial and marine) within a few years, so these other reservoirs would also have contributed their share to the carbon used for reforestation. These contributions reduce the amount of CO$_2$ required from the atmosphere. Allowing for all these factors, I found that reforestation after each pandemic would have been large enough to account for the observed CO$_2$ decreases of 4–10 parts per million.

It also occurred to me that pandemics could have caused a drop in the concentration of CO_2 in the atmosphere by slowing the global-average rate of deforestation. Although CO_2 was constantly being added to the atmosphere by deforestation over many millennia, it was also constantly being removed by uptake in the ocean. The processes that transfer CO_2 into the deep ocean work far more slowly than the changes in the rate of deforestation caused by a major pandemic. If a pandemic abruptly reduced the rate of CO_2 input, but the rate of removal to the ocean slowed only a little, the total amount of CO_2 in the atmosphere would have to drop.

Major pandemics are likely to have brought deforestation to a complete halt in the afflicted areas, based on historical evidence that reforestation actually occurred in such regions instead. These abrupt halts to deforestation across the stricken regions would then overwhelm the slow, steady deforestation still occurring in parts of the world spared by the pandemic. As a result, the global-mean rate of deforestation would drop, and atmospheric CO_2 concentrations would follow.

Humans could also have contributed to the CO_2 drops in a third way. In north-central China, where deforestation had occurred more than 2,000 years ago, most people are thought to have burned coal for heat and cooking. The massive levels of mortality during the Roman and medieval eras would have reduced CO_2 emissions from coal burning by amounts roughly proportional to the population losses in those regions. All of these factors—reforestation, reduced deforestation, and decreased coal burning—would have reduced atmospheric CO_2 concentrations. If natural (solar-volcanic) changes cannot provide an explanation, humans must have been an important factor.

One implication of these results is that CO_2 changes driven by pandemics could have played a significant role in climatic variations over the last 2,000 years. For a climate-system sensitivity of 2.5°C to CO_2 doubling, pandemic-driven reductions of CO_2 levels by 4 to 10 parts per million would have cooled global climate by 0.04 to 0.1°C. Such coolings would represent a significant fraction of the temperature changes observed (see fig. 12.3) between the cooler Roman era (200–600), the warmer medieval era (900–1200), and the cooler Little Ice Age interval (1300–1900). None of this rules out a role for solar-volcanic factors in these temperature changes, but the pandemic idea does manage to explain the size of the CO_2 reductions without violating the reconstructed temperature decreases, whereas natural solar-volcanic processes did not pass that test.

Part 3 of this book ended with the conclusion that a small glaciation should have begun several thousand years ago but was averted by a warming caused by slow increases in CO_2 and methane emissions by humans. For CO_2, the total anomaly caused by humans had grown to as much as 40 parts per million by medieval times. Then, during the Middle Ages, pandemics (perhaps aided by solar-volcanic

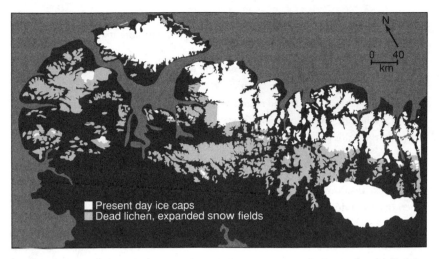

14.1. Surrounding small ice caps on Baffin Island that are rapidly melting today are halos of dead lichen that were smothered by the thick year-round snow cover prior to the 1900s.

changes) caused the CO_2 level to drop by as much as 10 parts per million. In effect, the pandemics eliminated a substantial fraction of the human-produced greenhouse "defense" that had been holding off a renewed glaciation. This abrupt CO_2 drop brought the climate system closer to the point of that long-overdue glaciation.

Indications of how close northeastern Canada may have come to being glaciated a few hundred years ago can be found on Baffin Island. There, as noted in chapter 9, several small ice caps today lie perched on the high plateau terrain. Surrounding these ice caps are large halos of dead lichen (fig. 14.1). These lichen halos are ghostly remnants of an ice age that nearly got under way during the Middle Ages.

Lichen are thin, green-gray blobs that grow on the surfaces of trees and rocks. Like other vegetation, lichen need sunlight, CO_2, and nutrients to power photosynthesis and permit growth. The Sun provides the needed light, and CO_2 is readily available from the air. But in contrast to most other forms of vegetation, lichen obtain nutrients from an unusual source: the seemingly inhospitable surfaces on which they grow. The lichen attack the rock and break it down into its component minerals and elements, which they then use as nutrients. Arctic lichen are also well adapted to extreme cold, since winter nights are often many tens of degrees below zero. As a result, extreme cold cannot kill the high-Arctic species.

Yet something did kill those areas of lichen and produce those ghostly halos. The only plausible explanation seems to be that the lichen were denied sunlight, and the best way to cut off sunlight is to bury them under thick piles of snow that are replenished in winter and never entirely melt in summer. So the dead-lichen halos are thought to be remnants from a time when permanent snowfields covered large areas of high terrain on Baffin Island.

Living on these same plateau crests today are younger lichen that have grown within the last 100 years. The interval during which these newer lichens have been growing corresponds to the recent warming since the end of the Little Ice Age. Taken together, the evidence tells this story: lichen once grew in the regions where now only the halos exist, then they were killed by the accumulation of permanent snowfields, and still later, about 100 years ago, the snowfields began to melt and new lichen began to grow. Decades ago, this evidence led scientists to conclude that the expansion of permanent snowfields that killed the earlier generation of lichen occurred during the Little Ice Age.

When snow persists from year to year in permanent snowfields, it gradually turns to granular snow (firn) and then to ice because of the pressure of the overlying snow and other processes. At first, the snow and firn and ice simply pile up in place but do not move. Only when the thickness of ice reaches several tens of meters does it begin to flow, at which point it can be called a glacier. As far as is known, the snowfields (or ice fields) atop Baffin Island during the Little Ice Age never thickened enough to flow, so technically they were not glaciers and this was not the start of a glaciation. But they were obviously a significant step in that direction.

In the same sense, the Little Ice Age was not really an ice age, just a time of slightly cooler climate. Still, in that one region of northeastern Canada, a first small step in the direction of a real ice age did occur, for a few centuries. I concluded that this small step toward glaciation had resulted in part from the drops in atmospheric CO_2 values caused by the American pandemic after 1500, combined with natural solar-volcanic changes (fig. 14.2). If such a small CO_2 drop (10 parts per million) had helped bring this region so close to a state of glaciation, then it seems likely that a significantly larger portion of northeastern Canada would have been glaciated if the remaining CO_2 anomaly as well as the methane anomaly caused by humans were removed from the atmosphere.

The pandemic/CO_2/climate connection has other implications. As noted in chapter 13, some climate scientists and historians have hypothesized that intervals of cold climate produce famines that cause population decreases, at least in highly vulnerable regions of the far North. During times of cold, wet weather, crops freeze or rot in soggy fields, and people starve. Cold has also been suggested as a cause of disease and death in populations weakened by hunger.

Recently, the link between human populations and climate during the last 2,000 years has been interpreted in a more provocative way. Based on the observation that global populations have tended to expand during warmer climates and contract during colder ones, the conclusion has been drawn that warm climates

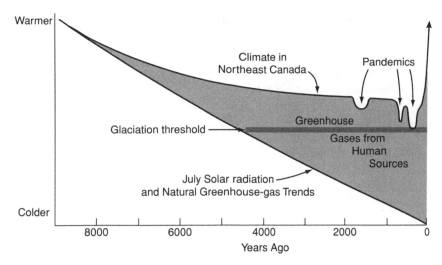

14.2. CO$_2$ drops caused by major pandemics probably brought parts of northeastern Canada to the threshold required for renewed glaciation.

must in general be good for humanity, and that cold climates must be bad. This conclusion has been injected into the global warming debate: any future greenhouse warming must be a good thing for humanity.

But does such a link between climate and population really make sense on a global scale? Historians have generally been resistant to this form of "environmental determinism"—the idea that climate is a major control on human population. One of their criticisms is that even the extreme decades of the Little Ice Age are thought to have been no more than 1–2°C cooler in winter, and only a few tenths of a degree cooler during the summer growing season. A second problem is that temperature drops of this size occurred only in far-northern and high-altitude regions, while subtropical and tropical latitudes and lower-elevation regions cooled even less.

Very few people actually live at the high altitudes and latitudes where the larger coolings occur, because agriculture in such places is marginal during the best of times. As a result, crop failures and famines in those regions have little or no impact on a continental or global scale. Most of humanity lives in the tropics and moist regions of the subtropics and has done so for millennia. To those at warmer latitudes, a small temperature decrease might well be welcome, if a cooling of a few tenths of a degree centigrade were even noticed. The major part of the world's population would seemingly not be impacted by a Little Ice Age cooling in any significant way.

Still, the data (see fig. 13.1) do suggest some correlation between global population and climate. Even though "correlation is not causality," this apparent correlation invites an explanation. My findings provide a possible answer: both the cooler intervals and the simultaneous population losses are independent responses to pandemics. Pandemics cause massive population losses for obvious reasons, but they also cause CO_2 reductions that contribute to cooler climates. In short, population losses do indeed correlate with cold intervals, but they are not caused by them. The common causal factor here is disease.

The connection between pandemics and CO_2 is also relevant to studies of estimated carbon emissions during the early part of the industrial era. Reconstructions of carbon emissions caused by land-use changes extend back to AD 1840, with cruder extrapolations back to about 1800. The current interpretation of the trends during the late 1700s and early 1800s is that land clearance had begun to increase early in the industrial era but had not yet accelerated as abruptly as it would after 1840.

The pandemic/CO_2 hypothesis adds a new factor to consider in these reconstructions. The American pandemic was still in full effect when the industrial era began because indigenous populations had never recovered from the decimation caused by European diseases. As European settlement began, huge expanses of forest were cut, usually beginning with the prime bottom land in and near fertile river floodplains. Almost all of these areas were the same regions that had once been cleared, densely occupied, and heavily farmed by native Americans but had reverted to forest after being abandoned. By implication, the rather high rate of deforestation and carbon emission in the Americas during the 1700s and 1800s was not entirely the result of a first cutting of virgin forest; it resulted in part from a *recutting* of forests that had initially been removed much earlier by indigenous peoples but had later grown back as millions of people died. If so, some of the clearance that has been included as part of the early industrial-era total actually represents a recovery from the impact of the American pandemic.

Finally, the results summarized in parts 3 and 4 of this book greatly complicate studies of climatic changes over the last several millennia. Until now, one of the primary reasons for studying these earlier intervals has been to define the natural behavior of the climate system prior to the onset of significant human impacts that were assumed to have begun 200 years ago. The goal has been to quantify the natural behavior of the climate system accurately so that we can isolate and clarify the effects of humans on the planet. But now it seems that no part of the last several thousand years of climate change is actually free from potentially significant human impacts. Good reasons still exist for studies of recent millennia and centuries, but the more important (and more difficult) task may now be to separate human impacts on climate from natural variations.

PART FIVE

Humans in Control

HUMANS IN CONTROL

DURING THE LATE 1700s and 1800s, the world entered the industrial era. Technological innovations such as engines powered by steam and later by gasoline transformed our ability to produce and make use of inexpensive power. The invention of the reaper and other devices transformed agriculture, enabling farmers to turn the deep roots of prairie grasses and plow the fertile soils. The population of the planet exploded from 1 billion in 1850 to 6 billion by 2000.

The environmental consequences of this era of innovation have been startling. By some accounts, humans now move more rock and soil than all of nature's forces (water, ice, wind, and landslides) combined. Croplands now occupy some 10 to 15 percent of the world's land surface and pastures take up another 6 to 8 percent, for a total of roughly 20 percent. Much of the useable land—those parts that are neither desert nor mountain nor tundra nor ice sheets—is now used for agriculture. More than half of the freshwater drainage is now in use by humans, most of it for irrigation. In most industrialized countries, nearly every acre of forest has been cut at one time or another. The natural flow of most of the world's rivers has been checked by dams, in many cases several times along each river. In all, some 30 to 50 percent of the land surface has now been significantly transformed by human action, with smaller changes in other regions.

Humans have now become the major environmental force on Earth. Nature can wrest back control on a regional scale and for a short interval, with a hurricane, a flood, an earthquake, a tornado, or a drought. But once the natural threat has receded, out come the bulldozers and the earthmovers to put things back the way humans want them.

In part 3 of this book, I made the case that humans took control of greenhouse-gas trends thousands of years ago, and that the gradual increases in gas concentrations prevented most of a natural cooling that would otherwise have occurred. In this contest, nature still exerted a slightly stronger control on climate than did humans. During the last two centuries, however, the human influence on greenhouse gases has increased markedly, with gas concentrations soaring to levels well beyond the natural range of the last 400,000 years, and the human influence on climate has become considerably larger than it was during the preindustrial era. But has it exceeded that of nature?

Most scientists agree about two observations: (1) greenhouse-gas concentrations have quickly risen above their natural levels during the last 200 years; and (2) global temperature has increased by about 0.6° to 0.7°C in the last 125 years at an unusually rapid rate. The immediate temptation is to draw the conclusion that an unprecedented warming caused by humans is under way.

Yet this conclusion is not so easily justified as it initially sounds. Any credible climate scientist will agree that at least some of the observed warming must be due to the measured rise in greenhouse-gas concentrations, but almost none would say that the *entire* warming has been caused *only* by greenhouse gases. Other factors also affect climate, and it is necessary to disentangle them before coming to a conclusion about the size of the industrial-era impact of humans on climate.

One seemingly anomalous observation is that the climatic warming of the last 200 years has been no larger than that caused by humans during the preindustrial era, even though the man-made increases of CO_2 and methane have been larger (chapter 15). The primary explanation is that the climate system takes decades to adjust fully to rapidly introduced greenhouse gases, and global temperature has not caught up to the explosive increase in these gases during the last half-century. Another likely factor is a partial cancellation of greenhouse-gas warming by other industrial emissions to the atmosphere.

The likely size of the future greenhouse warming depends on how much fossil carbon is economically accessible, how much of it we actually put into the atmosphere (and how fast), and how sensitive the climate system is to that input (chapter 16). A major uncertainty in such projections is whether or not future technological advances will reduce our greenhouse-gas emissions.

For millions of years, our human ancestors had no effect on climate. Then for several thousand years, we had a small but growing impact. In the last century, our impact has exceeded that of nature, and it will likely do so for many centuries to come. A millennium or so from now, the fossil-fuel era will largely have ended, and the climate system will be slowly returning toward its natural (cooler) state (chapter 17).

Chapter Fifteen

GREENHOUSE WARMING: TORTOISE AND HARE

READERS WITH A SHARP EYE may have been puzzled by a part of chapter 10 that seemed contradictory. Figure 10.1 showed that the relatively modest rise of greenhouse gases estimated to have been caused by humans before the Industrial Revolution (40 parts per million for CO_2 and 250 parts per billion for methane) led to a relatively large rise in global temperature (0.8°C), while the larger industrial-era rise in gases (now almost 100 parts per million for CO_2 and 1,000 parts per billion for methane) seem to have resulted in a global temperature increase of only 0.6°C. Although the relative sizes of these responses seem inconsistent, they are not. Two factors account for the differences.

The first (and primary) explanation lies in the tortoise-versus-hare difference between the two increases. Unlike the slow preindustrial rise in gas concentrations, the industrial-era increases have been so rapid that the climate system simply has not yet had time to register its full temperature response. The climate system needs decades to register its full response to any kind of push in a new direction, such as the addition of greenhouse gases. This delay is called the "response time" of the climate system.

Imagine that you have gone off on a winter vacation and have left a hot tub sitting outside. Not wanting to pay for unused electricity, you turn the temperature down and leave. When you return home, you decide you want to take a soak, so you turn the temperature setting back up, but an hour or more passes before the water heats up to the temperature you chose. That delay is a measure of the response time of the tub. Scientists would define the response time as the length of time the tub takes to heat up a certain fraction of the way toward the temperature setting you selected. A handy measure (the one used here) is the time for the tub to get halfway from the temperature at which it starts to the temperature setting you selected.

The effects of climate-system response times are in evidence in everyday life. Each day, the Sun's rays reach maximum intensity in midday, but the hottest temperatures occur hours later, in the late afternoon. Each year, radiation from the Sun reaches maximum intensity on June 21 in the Northern Hemisphere, but summer temperatures on land don't reach their maximum until mid-July or later. Both these daily and seasonal lags show that the climate system doesn't respond immediately; its response has a built-in delay.

TABLE 15.1
Response Times within Earth's Climate System

Climate-System Component	Response Time
Land	
Ice-free land	Hours to months
Ice sheets	Millennia
Ocean	
Low-latitude surface layer	Months to years
Polar surface layer	Months to decades
Sea ice	Months to decades
Deep ocean	Centuries
Global average	Decades

The problem for climate scientists is to determine the average response time of the entire climate system to the recent increases in greenhouse gases. But this is not easy: the climate system consists of many different parts, each with its own unique response (table 15.1). The problem is to allow for all these separate responses and figure out one average overall number for the planet as a whole.

If you were orbiting Earth out in space and looking down, you would see the major divisions of its surface: the land (30%) and the ocean (70%), with the land further divided into the fraction that is ice-free (27%) and the part covered by ice sheets (3%), and the ocean divided into ice-free areas (66%) and areas covered by seasonally fluctuating sea ice (4%). So one approach would be to find the response time of each of these separate parts of the climate system, weight its importance by the fraction of Earth's surface it covers, and come up with a weighted-mean average value incorporating all of the responses on the planet.

Ice-free land is the most familiar in daily life. Here in western Virginia well away from the ocean, the days can be as much as 25°C (40°F) warmer on a spring afternoon than on a frosty night. Part of the reason for this strong response is that heat doesn't penetrate very far into soil and rock, and the surface layer gets very hot as it absorbs the Sun's radiation. The fact that such a strong heating response can occur even within just a single afternoon is a clear indication that the response times of the land surface and the layer of atmosphere above it are short. For this reason, the interiors of continents in the Northern Hemisphere reach their maximum temperatures in July, only about a month after the maximum solar heating at the summer solstice. Based on all this information, the land and the air above it have a full response time to imposed changes in climate measured in just days to weeks.

Ice sheets cover a small part of Earth's land surface, and their behavior lies at the exact opposite end of the spectrum of climate-system responses (chapter 4).

Large ice sheets take thousands of years to melt, as shown by the most recent deglaciation that started near 16,000 years ago and finally ended about 6,000 years ago. And they grow even more slowly than they melt. The response time of ice sheets lies somewhere in the range of 5,000 to 10,000 years, the slowest rate in the entire climate system. The much smaller bodies of ice that occur as ice caps on mountain peaks or glaciers in mountain valleys are a different case—they react to imposed changes much faster, in just tens of years, but they cover a trivial percentage of Earth's surface and have little impact on Earth's average overall response.

So Earth's land surface area is dominated by regions with a very fast response but also includes ice-covered areas with very slow responses. Yet land accounts for only 30 percent of Earth's surface area. Most of the overall response of the planet must be determined by the oceans, which cover the other 70 percent. Oceans are also critical for another reason: water has the natural property of being able to hold large amounts of heat, and the oceans consequently are by far the climate system's largest storehouse of heat from the Sun.

Figuring out the response time of the ocean is complicated. Its surface layer (down to depths of a few tens of meters) heats up and cools off seasonally in response to changes in the amount of solar radiation that arrives at its surface and penetrates below. These seasonal swings in surface ocean temperature are much smaller than those on land, and they are delayed by one to two months. This delay explains why vacationers who visit the seaside in September find summer-like ocean temperatures even though the Sun's rays have been weakening for three months and air temperatures have cooled. The response time of ocean water is somewhat slower than that of the land, but much faster than that of ice sheets.

The surface ocean operates in different ways at low and high latitudes. In most of the tropics and subtropics, the lens of warm water in its uppermost layers persists year round. As radiation from the Sun warms this layer, winds in the overlying atmosphere stir the water and mix the heat to greater depths. Normal breezes stir the ocean only a little, but larger storms can churn it to depths of 50 to 100 meters (150 to 300 feet). In this way, the warmth of the uppermost layers is distributed downward into this "wind-mixed layer," and slightly cooler water from below is brought toward the surface. Because the strongest storms occur infrequently, it takes time for the heat to mix deeply. All of these processes figure into the response time of the tropical ocean. Estimates are that the full response time is about two decades, much longer than for the land, but much shorter than for the ice sheets.

Nearer the poles, heating by the Sun is much weaker, and no permanent lens of warm water exists, just a very thin layer of slightly less frigid water in midsummer. At these latitudes, the very cold temperatures during typical polar winters chill the salty ocean water and increase its density enough to send it hundreds of meters or more into the deeper ocean. This sinking water then flows from polar and near-polar seas toward lower latitudes.

This vast subsurface part of the ocean is one of the slow-responding parts of the climate system. The average parcel of water that sinks at the poles takes about 1,000 years to complete its journey through the deep ocean. Because the deep ocean accounts for over 90 percent of the total volume of the ocean, it might at first seem that the slow 1,000-year turnover of the deep ocean would dominate the overall ocean response.

In fact, however, the deep ocean is so isolated from the surface that its slow response doesn't effectively control the overall ocean response. Most of the water that sinks and flows away doesn't return to the surface for so long a time that the surface layers don't "feel" these slow changes occurring below.

Yet the high-latitude seas do figure into the overall ocean response time in several critical ways. First, part of the deep overturning at high latitudes that sends strongly chilled surface waters down also brings deeper waters back to the surface. Because this deep overturning occurs only in occasional winters years or decades apart, it takes decades for climatic changes at the surface to be distributed through hundreds or thousands of meters of ocean water. As a result, the response time of the surface ocean in small polar regions is slow.

One aspect of the high-latitude response that is particularly difficult to predict is whether or not a major warming will slow the rate of ocean overturning. Warm water tends to remain at the surface of the ocean because it is lower in density than colder water. The more a general climatic warming heats (or fails to chill) the surface-ocean layer, the greater its tendency will be to stay at the surface. In addition, a warming will also reduce the intensity of frigid outbreaks of cold air that flow out across the ocean in winter, extract its heat, and send chilled dense water to great depths. Any reduction of winter chilling should also weaken the overturning and leave warmer ocean water at the surface. Any broad reduction of overturning at near-polar latitudes would amplify Earth's response to a greenhouse-induced warming.

Sea ice is the final part of Earth's surface cover that affects the ocean response time. Around Antarctica, sea ice forms each winter in a layer 1 meter thick, but almost all of it melts the following summer. The response time of this thin, single-year ice is obviously fast. In contrast, much of the Arctic Ocean is covered by a layer 3 to 4 meters thick formed over several years, and most of that ice does not melt or even thin significantly from year to year. The response time of this thicker and more stable layer of Arctic ice is longer, probably in the range of years to a few decades.

When all of these influences on ocean circulation are considered together, estimates of the average response time of the ocean fall in a range between 25 and 75 years. Because 70 percent of Earth is covered by ocean, and because the oceans can store much more heat than the land, the average response time of the full climate system is in the same range, with a best estimate of perhaps 30 to 50 years.

This response time has different consequences for the two intervals of greenhouse-gas buildup. The preindustrial increases in CO_2 and methane were extremely slow, spanning thousands of years. Even though the full response of the climate system lagged decades behind the greenhouse-gas changes, the increases in gas concentrations from century to century were so small that the climate system was always close to full equilibrium with the then-current inventory of gases (except during the onset of the plague episodes discussed in chapter 14).

In contrast, the growth of greenhouse-gas concentrations in the atmosphere during the industrial era has been explosive (fig. 15.1). Well over half of these increases have occurred within my lifetime. The first hints of the industrial-era rise show up near 1800, but less than 20 percent of the increase had occurred by 1900. Even by 1950, less than 30 percent of the CO_2 and methane increases to date (as of 2005) had yet been registered. Over 70 percent of the industrial-era greenhouse-gas increases have thus occurred within an interval comparable to the estimated response time of the climate system.

As a result, the climate system has not yet had time to register a significant portion of the warming that will eventually occur. By some estimates, as much as half of the greenhouse warming that will eventually occur because of current greenhouse-gas levels is still in the pipeline. Even if we were somehow able to limit our future greenhouse emissions in such a way as to keep the concentrations of these gases in the atmosphere exactly constant for the next several decades, Earth's climate would still continue to warm as the climate system gradually came to its full equilibrium response to the current gas concentrations. This delay in registering the full climatic response is the main reason that the amount of warming in the industrial era seems too small relative to the gas increases, compared to the preindustrial changes.

A second explanation for this apparent discrepancy lies in other kinds of industrial-era emissions to the atmosphere that have cooled climate and counteracted some of the greenhouse warming. In contrast, the preindustrial era did not produce such emissions, or at least not in amounts sufficient to affect climate. The major emission from industrial smokestacks other than greenhouse gases is sulfur dioxide (SO_2), a gas sent into the atmosphere and transformed into tiny sulfate particles called aerosols. Unlike sulfur from volcanic explosions, these particles do not reach the stratosphere, where they could remain for a few years before settling out. Instead, they rise to heights of a few hundred or thousand meters and slowly drift away in the prevailing winds from the points of emission. The three major sources are the midwestern United States, Europe (especially eastern

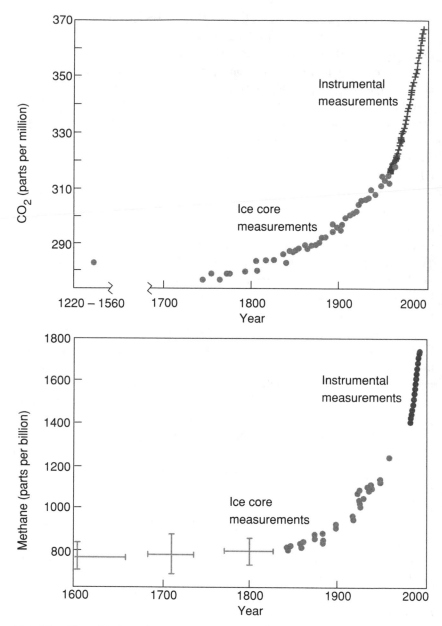

15.1. The effect of industrialization is evident in the rapid rise of atmospheric CO_2 and methane concentrations since the 1800s.

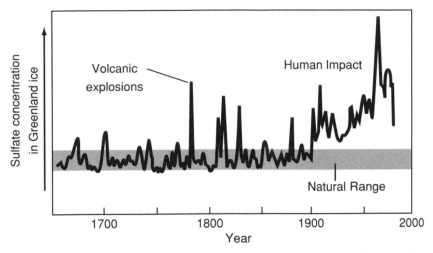

15.2. Measurements in Greenland ice show a rapid rise in atmospheric sulfate emitted by industrial-era smokestacks, followed by a downturn in the 1970s after passage of the Clean Air Act.

countries from the former Soviet Union), and China. Plumes of aerosols are located downwind (to the east) of these sources. Because these particles reflect some of the incoming solar radiation, their net effect on climate is thought to be a regional-scale cooling. The amount of cooling is not well known; it depends on intricate details linked to the size, shape, and color of the mixture of different kinds of particles, as well as their height in the atmosphere. Within days or weeks, these aerosols are scrubbed from the atmosphere by rain, but this may not happen until they have drifted hundreds or thousands of miles downwind.

Ice cores record part of the history of human emissions of aerosols. The sulfate content in an ice core from Greenland (fig. 15.2) reveals an interval from 1650 to almost 1900 when sulfate concentrations generally stayed within a low range, except for brief spikes that mark large volcanic eruptions in the Northern Hemisphere and the tropics. Then, around 1900, the background concentration began a steady rise marking the main part of the industrial era, especially in nearby North America. The downturn in the sulfate values around 1980 reflects the initial effects of the Clean Air Act in reducing SO_2 emissions in the United States.

Sulfate emissions during preindustrial times were probably negligible, at least compared to the industrial era. Neither Bronze Age nor early Iron Age metallurgy required extremely hot fires, and smokestack heights were low compared to those during the 1700s and afterwards. Both of these factors would have kept any SO_2 emissions close to the ground and within the source areas, rather than sending them higher in the atmosphere where they could be dispersed over large regions. These emissions are not likely to have cooled preindustrial climate significantly.

Sulfate particles are just one of a range of industrial-era aerosols we emit, and the impact of other aerosols on climate is even more uncertain. Dark particles of "black carbon" emitted from modern-day deforestation and other burning are thought to warm the lower atmosphere by absorbing solar radiation, but the size of the warming effect on global temperature is not known. Given that the rate of forest clearance and burning prior to the industrial era was only 5 to 10 percent as large as the modern-day rate, the slow clearance of forests prior to 1700 seems unlikely to have had any measurable effect on global climate because of the aerosols produced.

By some estimates, industrial-era aerosols could have cooled climate enough to cancel perhaps 20 percent of the industrial-era greenhouse-gas warming that would otherwise have occurred. This cooling effect is the second major part of the explanation invoked for the relatively small increase in global temperature during the last century or two despite the sizeable increase in greenhouse gases. Another factor that may yet figure in this story is the effect of industrial-era forest clearance in altering the reflectivity of Earth's surface and its response to solar radiation. In any case, the delayed response of the climate system and the cooling effects from sulfate aerosols are the most likely answer for the apparent discrepancy between the recent warming trends and the increases in greenhouse gases.

Paradoxically, any abrupt, all-out action to roll back both the greenhouse and smokestack emissions of the industrial era would have the effect of intensifying the greenhouse warming, rather than reducing it. If we stopped putting any SO_2 in the atmosphere today, the sulfate aerosols now there would rain out within a few weeks, removing their cooling effect and thereby causing some amount of warming almost immediately. But if we stopped putting CO_2 in the atmosphere today, it would take more than a century for half of the industrial-era CO_2 excess now in the atmosphere to be taken up in the deep ocean. With the industrial cooling effect quickly removed but the warming effect still in place, climate would warm. In addition, for at least a few decades, the unrealized greenhouse warming that is still in the pipeline in response to past greenhouse-gas increases would continue to kick in, pushing temperatures even higher. It's strange, but true: by cleaning up all our industrial emissions, we would amplify global warming, at least for a few decades. Later, as CO_2 levels gradually decreased and the climate system slowly responded, climate would begin to cool.

Chapter Sixteen

FUTURE WARMING: LARGE OR SMALL?

THE SIZE OF THE FUTURE greenhouse warming hinges mainly on two issues: (1) How high will the concentrations of CO_2 and equivalent greenhouse gases rise because of human activities? (2) How sensitive is the climate system to those increases? The answers to both of these questions have significant uncertainties attached.

It took millions of years for natural processes to store three carbon-based energy sources in Earth's sedimentary rocks. The carbon stored in coal originated in vegetation that rotted in oxygen-deficient swamp waters on low-lying continental areas. The carbon stored in oil and natural gas came from the organic remains of ocean plankton and other debris in shallow seas. Over time, the heat and pressure of burial transformed this raw organic matter into coal, oil, and gas. Most coal deposits are hundreds of millions of years old. Oil and gas are almost as old—a few tens to a few hundreds of million years in age.

In contrast, the modern era of dependency on carbon fuels is still in its infancy, having begun only in the mid-1800s. Initially, coal was the fuel of choice for industrial energy and for home heating, but oil and gas have since displaced coal for many purposes. Yet the ultimate limits to these resources are already beginning to come into view.

Oil is virtually the only fuel used in cars and trucks, but the supply is not unlimited. By some recent estimates (for example, Ken Deffeyes's book *Hubbert's Peak*), sometime between 2004 and 2009 we will reach the year marking the largest amount of oil pumped in Earth's history (or future). After this peak year passes, the amount pumped (and consumed) will slowly decrease for a century or so until most of the economically accessible oil is depleted. Even more optimistic projections suggest that oil production will decline not long after the current decade.

The year of greatest-ever oil production in the United States was 1970, and the decline in recovery since then has been steady. The huge oil fields of the American Southwest and West, once the largest producers in the world, had all been discovered by the late 1940s, and no new fields on a major scale have been found in that area since that time, despite intensive exploration and drilling. Using innovative new techniques, oil companies have located pockets of oil lying in between the large, tapped-out fields, but these finds have only kept the plunge in oil recovery over the last three decades from being even faster than it has been. These "new" sources, along with fields in Alaska, have slowed the growing U.S. reliance on foreign oil only marginally. Estimates are that 15 to 30 billion barrels of oil lie under Alaska's North Slope at Prudhoe Bay, but the world consumes that amount every four years.

The largest oil fields in the world by far are those of the Persian Gulf region of the Middle East, yet all of the major discoveries in that region had already been made by the 1950s. Some confusion exists about the size of the proven reserves in the Gulf region. In the1980s the Persian Gulf nations abruptly increased their estimated oil reserves at the same time, just after the Organization of Petroleum Exporting Countries (OPEC) decided to use the size of the reserves as a basis for allocating shares of annual production. Because no new finds were announced at that time (or since), many analysts reject those larger numbers as motivated by the short-term economic concerns of countries wishing to maintain their previous national share of total OPEC production.

If the earlier (smaller) estimates of Gulf oil reserves are correct, the peak production year is not far away. Much of the world's remaining oil reserves lies in smaller pools dispersed among the major fields already being drilled, similar to the historical pattern in the United States. Even if the larger reserve estimates from the Gulf regions prove to be correct, the peak production year is still not far off (perhaps two decades). The OPEC nations could choose to keep pumping their known reserves at an ever-faster rate, thus pushing the peak-production year farther into the future. But if they do so, the reserves will be depleted faster, and the decline in production will be much steeper when it occurs.

Some optimists remain certain that vast oil fields lie undiscovered in unexplored regions, but Deffeyes argues that few pleasant surprises await us because most regions on Earth have already been explored. First, geologists map surface rock outcrops, followed by geophysicists using seismic equipment to obtain images of the subsurface layering. Next come exploratory wells drilled to recover subsurface samples for examination by geochemists. Laboratory analyses of these samples indicate whether temperature and pressure conditions have been favorable for oil to form in each region. Because almost every region on Earth has by now had exploratory wells drilled, the chance that really huge undiscovered oil fields exist is not high. Because of ongoing territorial disputes, the South China Sea is the last major unexplored region on Earth, but geologic considerations make it unlikely that oil fields on the scale of the Persian Gulf lie there. Drilling ever deeper into already explored areas will not help: oil forms under conditions of moderate temperature and pressure at relatively shallow depths (about 1.5 to 5 kilometers beneath the surface).

Meanwhile, the global appetite for oil continues to grow. An oil field that would have provided many years of world supply a few decades ago will now last only a fraction of that time because demand has been going up relentlessly. As nations in eastern Asia continue to industrialize, consumption can only increase. At some point, the relentlessly rising curve of oil consumption must overtake the amount that can be economically pumped, and oil production must level off and then turn downward. Technically, we will never "run out" of oil. Somewhere in Earth, oil will exist that has not yet been drilled, and some of it will still be worth pumping. But extracting these ever-smaller and more dispersed sources will gradually become more costly. As the cost of extraction creeps up toward the economic reward from doing so, oil-producing companies will gradually abandon the effort, as many have in the United States.

For all of these reasons, the time of maximum annual production of oil (and peak CO_2 emissions to the atmosphere from oil use) is likely to arrive within a decade or so, initiating an era in which we become more dependent on natural gas and coal. Gas forms in the same regions as oil and from the same sources, but at higher temperatures and pressures. For many years, oil drillers simply let most natural gas burn off into the atmosphere at the wellhead. Now that gas is recognized as an energy source for industries and homes, it is valued as highly as oil. At current and projected rates of use, natural gas reserves will last somewhat longer than oil. In addition, because gas is stable at higher temperatures and pressures than oil, large sources may yet lie undiscovered deep in several regions, including Siberia. Nevertheless, estimates are that a world gradually switching from reliance on oil will consume most of the remaining gas reserves by the end of this century. With accessible oil and gas reserves largely depleted, coal will be the main fuel source left for the twenty-second century. The reserves of coal (mainly in Russia, China, and the United States) could provide energy for an additional century or two, but at a greater environmental price.

Until recent decades, extracting coal depended on hunched-over miners working narrow coal seams in hazardous and unhealthy conditions that are hard to imagine today. The work was an endless cycle of crouching or lying on damp mine floors, digging out the layer of rock lying just beneath the coal, blasting out a few feet of exposed coal, shoveling and hauling it out, and then going back and doing the same thing again. Now the work is mostly done by machines and on a much larger scale. Rather than tunneling in, enormous earth-moving machines simply remove everything lying above the coal seams. The coal seams are only a few feet in thickness and form just a small portion of the total amount of rock that must be moved. As a result, coal extraction produces an enormous volume of rock debris that has to be put somewhere. In West Virginia, the topography of large regions is being reconfigured to retrieve coal: mountain tops are shaved off and valleys are filled with rock debris, with varying levels of environmental safeguards. The costs in money and energy to retrieve coal in this way far exceed those for oil and gas but are still economical.

With current technology, coal emits more CO_2 per unit of useable energy yielded than oil or gas. The types of coal that provide the most energy per ton were mined earliest, and much of the coal that remains is less efficient and rich in sulfur, which accumulated in those ancient swamps along with the plant carbon. As coal begins to replace oil and gas as an energy source, both CO_2 and sulfur emissions will go up for each unit of energy used. Technological innovations may mitigate sulfur emissions from coal burning, but affordable ways to reduce CO_2 emissions are nowhere in sight. Based on what we know now, global CO_2 emissions will rise markedly as the world turns to coal for energy.

By 2004 the CO_2 concentration in the atmosphere had increased by 33 percent from the preindustrial level, from 280 to 375 parts per million. Other greenhouse-gas emissions from human sources (including methane, ozone, and nitrogen oxides) had also occurred, so that the total greenhouse-gas increase in the atmosphere was by then equivalent to a 50 percent increase of CO_2 (see fig. 15.1). The obvious question is how high this CO_2 trend will rise in the future.

Several factors, some inherently hard to predict, figure into attempts to project future CO_2 concentrations. The total amount of CO_2 emissions generated by consuming the world's reserves of fossil fuel is one variable, and the rate at which these reserves are used is another. It takes an average of about 125 years for the deep ocean to rid the atmosphere of each extra molecule of CO_2 added during the industrial era. If all the CO_2 is added quickly, the ocean's ability to absorb it is overwhelmed. If it is added more slowly, the ocean has more of a chance to re-move CO_2 and limit the peak concentration reached.

The largest uncertainty of all may be the effect of technological innovations on future emissions. It is not unreasonable to hope that human ingenuity will invent processes that will allow us to consume coal but not send so much CO_2 into the atmosphere. Still, disposing of hundreds of billions of tons of oxidized carbon will not be easy or cheap.

Despite these uncertainties, scientists and economists have constructed plausible scenarios to bracket the likely range of future trends (fig. 16.1). The more opti-mistic scenarios assume that humanity manages to moderate (reduce) greenhouse emissions by some combination of public policy, voluntary action, and techno-logical innovation. According to these scenarios, greenhouse-gas concentrations will rise to the $2 \times CO_2$ benchmark level near 2200 or possibly earlier, remain high for 100 years or so, and then slowly drop. Even this optimistic projection brings us to equivalent CO_2 levels not seen in Earth's atmosphere for at least the last 5 to 10 million years.

The second group of scenarios is often referred to as the "business as usual" projection. These scenarios assume neither major technological innovations nor strong political or public commitments to reducing greenhouse emissions. By

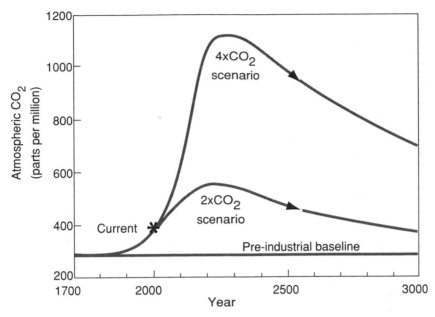

16.1. Atmospheric CO_2 concentrations are likely to reach two to four times the natural (preindustrial) level in the next few centuries as fossil-fuel carbon is consumed.

2200 to 2300, greenhouse gas concentrations rise to an equivalent CO_2 level at least four times higher than the preindustrial value, a level higher than any in the last 50 million years of Earth's history.

Which of these scenarios is more likely? No one really knows. Even if new technologies become available to reduce CO_2 emissions, less affluent countries may not be able to afford them, and affluent countries may resist them if they fear losing their economic competitive edge. As a result, it will be very difficult to avoid reaching at least the $2 \times CO_2$ level. Even if we were to freeze the rate of carbon emissions at their current levels, we would still reach the $2 \times CO_2$ level because the new CO_2 emitted per year is accumulating faster than the carbon already in the atmosphere can be absorbed into the ocean. The 1997 Kyoto emissions treaty attempted to limit future CO_2 emissions to late-1990s' levels, but the CO_2 concentration in the atmosphere will continue to rise at a fast rate even now that the treaty has been adopted. The momentum of modern industrialization seems likely to carry us to well beyond the $2 \times CO_2$ level.

The second major factor that will determine the size of the future greenhouse warming is the sensitivity of the climate system to a doubling of CO_2 (including equivalent contributions from the other greenhouse gases). I have cited several times the estimate that global temperature will increase by 2.5°C (4.5°F) if CO_2

levels double. That estimate comes from an international collection of scientists on the Intergovernmental Panel on Climate Change, and it lies near the middle of a range of estimates (1.5° to 4.5°C) based on many kinds of climate models. This range of possible sensitivity values is encouragingly narrow in one sense (at least they agree as to direction!) yet annoyingly large for those who want specific estimates of the future.

The largest uncertainties in the climate models are associated with clouds. High, wispy cirrus clouds warm climate by intercepting back-radiation emitted from Earth's warm surface and keeping it in the atmosphere. Lower, darker cumulonimbus clouds tend to cool climate by reflecting more incoming radiation from the Sun than the back-radiation they intercept from the Earth. And other types of clouds at different heights in the atmosphere play roles that are not well known as yet.

Although most models tend toward the middle of the range of $2 \times CO_2$ sensitivity estimates, some climate scientists disagree with this estimate. Some of those who infer a lower (~1.5°C) level of sensitivity point to the small size of the warming in the last century compared to the known increase in greenhouse gases. Such a comparison ignores the argument that the warming measured in recent decades is incomplete because of the delay in response of the climate system. This view also rejects any major cooling effect from aerosols (chapter 15). On the other hand, neither the size of the climate-system delay nor the climatic effect of the aerosols is well known at present.

For the midrange sensitivity estimate of 2.5°C, figure 16.2 shows the average global temperature changes that would result from greenhouse-gas increases to the $2 \times CO_2$ and $4 \times CO_2$ levels. For a tangible sense of what those amounts of warming would mean, we can look back in Earth's history to times when CO_2 levels were comparably high.

Earth last had CO_2 levels twice the preindustrial value some 5 to 20 million years ago, an interval with much less ice than now: no Greenland ice sheet, less Arctic sea ice, and no mountain glaciers, except perhaps in a few mountains in Scandinavia. In the Antarctic, sea ice was much less extensive and the vast polar ice sheet on the continent was somewhat smaller than today. The vegetation was also different in north polar regions: in place of the broad band of tundra and permafrost (permanently frozen ground) now surrounding the Arctic Ocean were conifer forests (spruce and tamarack) like those now found south of the tundra. Many kinds of vegetation lived at higher latitudes and altitudes than today because temperatures were warmer. This world was noticeably different from the one in which we live.

Is this the world our descendants will be living in two or three centuries from now? The answer to this question is neither yes nor no, but something in between.

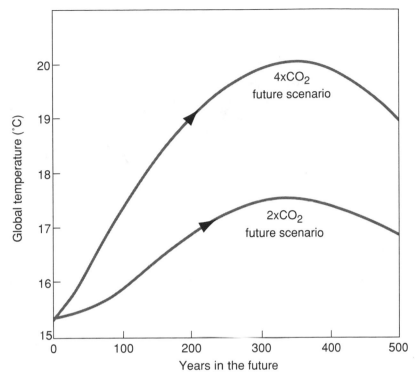

16.2. Global temperatures are likely to increase by several degrees Centigrade as CO_2 and other greenhouse gases are added to the atmosphere.

It depends on which part of the climate system you have in mind. The degree of reaction of each part of the climate system to the future warming will be determined by its response time (see table 15.1). Because the main part of the peak in greenhouse-gas concentrations and the resulting peak in warmth will last for about 200 years, the fast-responding parts of the climate system will be drastically altered, but the slow-responding parts will be much less affected.

At the slow-responding extreme, ice sheets take thousands of years to react to changes in climate. As a result, the lower and warmer margins of the Greenland ice sheet will begin to melt in this warmer climate of the future, but the bulk of the ice sheet may have been relatively little affected by the time the greenhouse-gas pulse begins to fade (although recent analyses point to greater vulnerability). The deep-frozen ice sheet on Antarctica will probably be little affected. Air masses there will become somewhat less frigid and thus able to hold and carry more moisture toward the pole. Rates of snow accumulation will increase on the high central portion of the ice sheet. On the other hand, along the lower ice-sheet

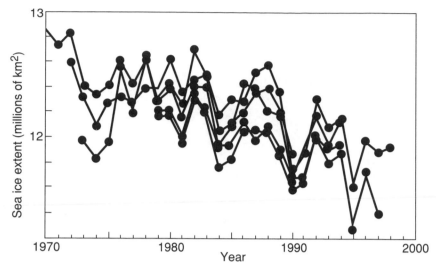

16.3. Several kinds of satellite-based measurements show that Arctic sea ice has retreated during the last several decades.

margins, a slightly warmer ocean may increase the removal of ice. Whether the Antarctic ice sheet as a whole will grow or shrink depends on which process wins this battle. Most likely, it will not change in size very much.

The situation is clearer for mountain glaciers. Because these small bodies of ice respond to climate change within decades, they will be heavily impacted by future warming, and most will disappear completely. At present, virtually every mountain glacier on Earth is already melting in response to the industrial-era warming. Within the last century, some 70 percent of the glaciers in Glacier National Park have disappeared (mostly the smaller ones). At current rates, the larger ones will disappear by 2030 from "Glacierless National Park."

A similar fate may be in store for sea ice, which also responds relatively quickly to changes in climate. Summer sea ice may completely disappear from the central Arctic Ocean, and winter sea ice may no longer reach the coasts of North America and northern Eurasia. Here again, a trend in this direction is currently under way. Several kinds of satellite measurements over the last three decades show a 6 percent retreat in the area of Arctic sea ice (fig. 16.3). More ominously, radar measurements from submarines transiting the Arctic Ocean show a 40 percent decrease in sea-ice thickness from a regional average of 3.2 meters (10 feet) in the 1960s to just 1.8 meters (6 feet) in the mid-1990s. If this thinning continues in future decades, rates of areal retreat will accelerate. Seasonal snow limits in the Northern Hemisphere are another of the fast-responding parts of the Arctic climate system, and satellite measurements show that the extent of Northern Hemisphere snow has decreased in the last three decades. These sea-ice and snow-cover trends might yet prove to be part of a natural cycle of oscillations, or they may represent the first steps in a longer-term Arctic thaw that will intensify in the next two centuries. Time will tell us which is true.

The ring of permafrost that surrounds the Arctic responds somewhat more slowly to climate. Most of the frozen ground lies too deep to be easily reached by the seasonal heat from the summer Sun. The future greenhouse warming will warm and melt the surface layer and turn it into a quagmire for a much longer interval of the year than at present, but it will not melt much of the deeper layer of frozen ground. Permafrost (and tundra) will also shrink or disappear from high-mountain regions farther south.

In temperate latitudes, the length of the growing season will increase, and with it the fraction of the year when forests and grasslands are green. On average, the warm season will expand by about a month at either end. Future Aprils will be like modern Mays, and future Novembers like modern Octobers. Again, such a shift is now under way: the growing season measured both by satellites and by ground observations has expanded by about a week in the spring and by half a week in the autumn over the last two decades. Part of this change probably reflects the greater availability of atmospheric CO_2 in giving plants a fertilizer boost, but part of it is also due to the warming in recent decades. Another significant change at temperate latitudes will be a reduced number of bitterly cold outbreaks of polar air masses in winter.

At tropical and subtropical latitudes, where future temperature increases will be smaller, the main concern will be drought and floods. As global temperature rises, evaporation will increase, because rates of evaporation are determined mainly by temperature. With more water vapor drawn into the atmosphere each year, more rain must fall. If the distribution of rain were perfectly even, the increase in precipitation in each area might balance the increase in evaporation, with little or no net effect. But rainfall is notoriously patchy, especially in the warmth of summer when thunderstorms let loose the most torrential rains of the year. As a result, the most likely outcome will be both more extensive droughts in some areas and more severe storms in others, with relatively little predictability as to location.

If the future greenhouse-gas concentrations and temperature trends follow the "business-as-usual" scenario and reach values four times the preindustrial concentration (figs. 16.1, 16.2), all of the trends just described will be intensified. A $4 \times CO_2$ level of greenhouse gases is equivalent to the levels that existed some 50 million years or more ago when no permanent ice existed anywhere on Earth, even in Antarctica. For this large a future warming, the present-day Antarctic ice sheet would be out of equilibrium and would join the Greenland ice sheet in shrinking. But once again, the peak centuries of the greenhouse interval would pass too quickly for a large fraction of the south polar ice to melt. Future owners of coastal property will thank the sluggish response of these ice sheets. If all the water locked up in the two ice sheets were released, sea level would rise by 66 meters (205 feet).

In summary, whether we eventually reach the $2 \times CO_2$ level, the $4 \times CO_2$ level, or (more likely) something in between, the future greenhouse warming will be large. Having inadvertently stopped a small glaciation from developing in the last several thousand years, we will now melt much of the world's sea ice and mountain glaciers in the next century or two and push back the seasonal limits of snow cover, but we will leave the two great ice sheets largely intact.

Our impacts on the climate system will change an important part of Earth's basic appearance from even the distant perspective of space: most of the north polar region that is now white will be repainted in hues of dark blue (where sea ice melts back) and dark green (where snow-covered tundra gives way to boreal conifer forest). As satellite photos accumulated over many decades begin to show that we are repainting Earth's northern pole, the scale of our impact on climate will become obvious to all.

Chapter Seventeen

FROM THE PAST INTO THE

DISTANT FUTURE

The role of humans in Earth's climatic history falls into four phases (fig. 17.1).

Phase 1 (before 8,000 Years Ago)

Until 8,000 years ago, nature was in control. Even though our remote prehuman precursors had been present on Earth for several million years, nature alone drove climate change. Even when our fully human ancestors appeared sometime after 150,000 years ago, our impact on the global landscape was still trivial. People used "firesticks" to burn grasslands or forested areas in order to drive game or provide open areas to attract game and permit the growth of berries and other natural sources of food. Some of these early cultures may have pushed small branches into moist soils in wet tropical regions where trees bearing fruit or nuts would naturally sprout.

Yet both the number of humans and our footprint on the landscape remained small and highly localized in scale. The impacts of human fires were largely inseparable from the effects of natural lightning strikes, which cause random fires of varying intensity. Humans who set fires in one region and then moved on to another as part of their hunter-gatherer lives mimicked natural events. The number of humans setting fires was not yet large enough to rise above the background level of fires occurring as a part of nature.

Nearer the present, in the millennia just before 11,000 years ago, large changes in Earth's climate were under way in response to natural causes. High summer solar radiation in the Northern Hemisphere was melting the huge ice sheet on North America and the smaller ones on Scandinavia and in far-northern Eurasia. The strong summer Sun that provided the basic driving force for this immense effort was aided by high (natural) values of two key greenhouse gases: carbon dioxide and methane. Together, the Sun and these two greenhouse gases managed to melt the great ice sheets of the Northern Hemisphere over an interval of 10,000 years. But these were the same natural processes that had been going on for the several million years of Northern Hemisphere ice-age cycles. Nature was still in full control of climate.

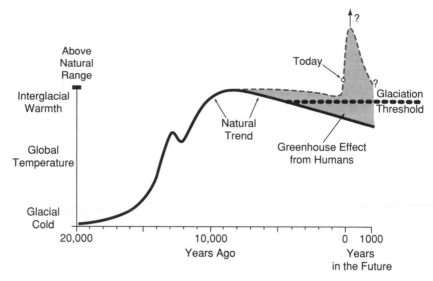

17.1. Greenhouse-gas emissions from farming have offset part of a natural cooling since 8,000, years ago and probably prevented a new glaciation. The rapid changes caused by modern industrialization will eventually reach levels of warmth not attained on Earth for many millions of years. Once the supply of fossil fuels is depleted in a few centuries, climate will gradually cool toward natural levels.

Ice sheets naturally respond very sluggishly to any push toward change: the word "glacial" is synonymous with "slow." As a result, even though the maximum push to melt the ice sheets occurred nearly 11,000 years ago, the ice sheets did not actually finish melting until 6,000 years ago. During the intervening 5,000 years, control of climate in the Northern Hemisphere became a contest between two great opposing forces: (1) the ice sheets, which had previously kept climate cold but were now rapidly melting, and (2) solar radiation and the greenhouse gases, which had reached peak (warm) values nearly11,000 years ago but were now beginning to decline toward lower modern values. The result of this contest of contending forces was a compromise: maximum temperatures were reached nearly8,000 years ago, when the ice sheets had been sufficiently reduced in size to have little cooling impact, but when solar radiation values were still high enough to make climate warm.

PHASE 2 (8,000 TO 200 YEARS AGO)

As the last remains of the great ice sheet in North America melted, humans began to clear forests in southern Europe and northern China to make way for agriculture roughly 8,000 years ago, and deforestation gradually spread to other parts of southern Eurasia over the next several thousand years. Burning of trees added CO_2 to the atmosphere at rates that initially were small but increased steadily through time. These carbon emissions reversed the natural decline in CO_2 levels that had occurred in the early parts of previous interglaciations. By clearing forests for farms, humans had begun to take control of Earth's greenhouse gases and in this way to exert a small but growing impact on Earth's climate.

Humans also began to irrigate lowland areas in Southeast Asia 5,000 years ago. The lowland areas flooded for rice crops added methane, another important greenhouse gas, to the atmosphere. As a result, a natural drop in methane values was reversed, and the atmospheric CH_4 trend began to rise instead of fall because it too had come under human control.

Forest clearance and irrigation for rice increased steadily through both the Bronze Age (which started nearly 6,000 years ago) and the Iron Age (which started nearly 3,000 years ago). When horses and oxen were domesticated and hitched to metal plows, each farmer could clear and cultivate larger amounts of land. Long before the start of the industrial era, most of the arable land along the southern tier of Eurasia had been cleared of forests, and most of the lowland delta regions of Asia were planted in rice.

By the start of the industrial era, human emissions of CO_2 and CH_4 had caused increases in the atmospheric concentrations of both gases equivalent to about half of the natural range of variation that had occurred previously. Greenhouse-gas concentrations in the atmosphere remained within their natural range of variation, but the levels were approaching the top of that range. These increases caused by humans produced a greenhouse warming effect that canceled a large portion of a natural cooling under way at high latitudes. This warming effect probably kept an ice sheet from beginning to grow in northeastern Canada during the last several thousand years. Humans had not yet taken control of climate (global temperature), but our effect on global climate was now very nearly equal to that of nature.

PHASE 3 (200 YEARS AGO UNTIL 200–300 YEARS IN THE FUTURE)

The industrial era arrived in the late 1700s and the middle 1800s, and it ushered in a third phase of human effects on climate. Rates of deforestation increased to provide fuel for mills and mining operations and to open up new farmland for rapidly increasing populations in tropical areas. Beginning in the late 1800s, use of fossil fuels (first coal, and later oil and natural gas) rapidly increased, eventually replacing deforestation as the primary source of CO_2 emissions by humans.

Methane emissions also accelerated, in part because of continuing increases in ir-rigated areas, but more importantly from the spread of methane-emitting land-fills, releases of natural gas, and other human activities.

Since the 1800s, atmospheric concentrations of carbon dioxide and methane have risen exponentially. Late in the 1900s, concentrations of both gases reached levels not recorded in ice cores for the last several hundred thousand years, and thought to have last existed on Earth many millions of years ago. At the start of the current millennium, CO_2 concentrations were about 30% higher than the levels typical of previous natural interglaciations and were increasing at a rate of about one-half percent per year. Methane concentrations have now reached a level almost 2.5 times higher than those during previous interglaciations, with in-creases of about 2% per year. Each new increment of CO_2 and methane carries us farther into uncharted greenhouse territory.

Global climate has warmed by 0.6° to 0.7°C during the last 125 years, the in-terval for which enough ground stations exist to make reasonably valid estimates of global temperature. Even though the greenhouse-gas levels have risen well beyond natural levels, the global temperature value reached by the start of the current millennium has not yet exceeded those reached during previous inter-glaciations of the last several hundred thousand years.

The explanation for this seeming discrepancy lies in several factors covered earlier:

1. Solar radiation values in the Northern Hemisphere during the last several thousand years have fallen well toward their minimum natural levels as changes in Earth's orbit have become favorable for renewed glaciation (chapter 10). As a result, a large part of the unprecedented rise in green-house gases caused by humans has been used to offset an overdue glaciation.

2. The decades-long response of the climate system has delayed a substan-tial part of the warming that will eventually result from the current level of greenhouse gases (chapter 15). When the climate system comes to its full response to the current greenhouse-gas levels, Earth will likely enter an era of warmth unprecedented during the previous millions of years of glacial-interglacial cycles. Of course, by that time, greenhouse-gas levels in the atmosphere will have risen even higher, and Earth's temperature will be moving toward an even higher value.

3. Industrial-era emissions of sulfate aerosols have probably canceled part of the warming that greenhouse-gas emissions would otherwise have caused (chapter 15).

At the current rates of increase of greenhouse-gas concentrations, future tem-peratures are likely to be pushed beyond the natural glacial-interglacial range of variation in the very near future, probably within a decade or two. Over subsequent decades, as greenhouse-gas concentrations rise at unprecedented rates to levels unprecedented in the last several million years, increases in global temperature will follow as the climate system responds. Deforestation in the tropics will play a significant role in the CO_2 increase, but by far the largest factor will be the burning of the world's remaining oil, gas, and (especially) coal reserves. Predictions of gas concentrations and global temperatures more than a century from now are impossible, but what we know now indicates that the increases will be large. The major hope for preventing high CO_2 levels lies in developing new technologies for trapping carbon before it leaves smokestacks or exhaust pipes. But for now, this is only a hope; no technology that would be economically feasible on the massive scale required is now in sight.

Phase 4: After 200–300 Years from Now

A few centuries from now, well after the economically recoverable reserves of oil and natural gas are largely depleted, and as the supply of accessible coal is also in decline, the diminishing rate of CO_2 emissions to the atmosphere from human consumption of fossil fuels will eventually fall below the rate at which the ocean takes up the excess carbon we have produced. At that point, we will enter a new phase in the relationship between humans and climate: the CO_2 concentration in the atmosphere will begin to fall, and eventually much of the industrial-era CO_2 excess we have caused will be removed.

In this longer run, the CO_2 taken up by the ocean will take part in a chemistry experiment far beyond anything attempted in any laboratory. The extra CO_2 will make the ocean slightly more acidic than it is today, and this acidity will dissolve some of the soft, chalky sediments lying on the sea floor. These chalky "oozes" are formed from the calcium carbonate ($CaCO_3$) shells of plankton that live in shallow ocean waters and settle to the sea floor after they die. The deepest ocean waters, below about 4,000 meters, are corrosive and dissolve most of the shells raining down from above. The deeper ocean basins are left with a brownish ($CaCO_3$-poor) residue of debris blown out to sea from the continents, while whiteish ($CaCO_3$-rich) sediments accumulate only on higher elevation features like snow on mountain peaks. As the excess CO_2 generated by humans makes ocean waters more corrosive, they will begin to attack existing deposits of $CaCO_3$ ooze on the higher sea-floor topography, and the oceanic $CaCO_3$ "snowline" will retreat to shallower levels. In this way, the CO_2 generated by humans will slowly be consumed in a gigantic chemistry experiment—dissolving $CaCO_3$ on the sea floor.

Meanwhile, as the ocean slowly takes up the CO_2 excess, global temperature will gradually cool back toward natural levels. If the greenhouse-gas concentrations fell all the way back to their natural level, Earth would probably cool enough to allow an ice sheet to begin accumulating in northeastern Canada, a return to an overdue glaciation. But this will not happen, or at least not for many millennia.

A small fraction—perhaps 15 percent—of the CO_2 generated during the preindustrial and industrial eras will remain in the atmosphere for several thousand years and keep global temperatures warmer than the natural value. In addition, methane concentrations will remain high as long as we continue irrigating for rice farming and storing our waste products in large, methane-emitting landfills. These activities will produce large amounts of methane every year and keep atmospheric concentrations well above natural levels. It is even possible that we will never return to that now-overdue glaciation.

But of course, no one can see centuries into the future.

Epilogue

EPILOGUE

"GLOBAL CHANGE"—a term that encompasses the size and impact of future climate change caused by greenhouse-gas emissions from human activity—is one of the most highly polarized topics in all of science. Scientific assessments of this impact have direct implications for key economic areas such as transportation, electricity generation, and heating and air conditioning. Large amounts of money hinge on governmental decisions about what energy policy to adopt in response to the problem, and this link makes global change a major political issue.

In the debate over global change, environmentalists tend to be arrayed on one side of the issue, concerned about damage to the environment if future fossil-fuel use causes large-scale climate change. Some industries take the opposing position that efforts to mitigate the impact of global change would be damaging to the economy. Numerous exceptions to these generalities can be found within both groups.

Because of this polarization, several disclaimers on my part are in order. I have never published an opinion piece on the global-change issue. I have never received any funding from either environmental or industry sources. All of my career funding has been from the government, and over 99 percent of it from the National Science Foundation, which is widely regarded by politicians of many views as the model of a well-run government funding agency (based on its reliance on competition and peer review). All of the funds used to write this book came from my retirement annuity earned at educational institutions.

Most of my research has been on climatic changes in the distant past, including the ice-age cycles of the last few million years, and the long-term cooling of the last tens of millions of years. I did address the global-change issue in a college-level textbook published in 2001. One reviewer cited that treatment as a model of balance, and the textbook has drawn no published criticism that I know of from either the environmental or industry extremes. In sum, I have no prior public record as an advocate on either side of this issue, and no financial interest in the outcome of the debate.

Nevertheless, I do have opinions on global change. Given the polarization around the current debate, I have chosen to isolate these opinions in this "editorial" section.

Distortions come from both extremes of the global-change debate (chapter 18). Environmental extremists are mostly prone to alarmist exaggerations, while pro-industry extremists systematically attack or even deny basic knowledge coming from mainstream science. In my opinion, these trends are reaching the point where they may do damage to the integrity of climate-science research.

Placed within the larger framework of environmental and resource concerns (chapter 19), global climate change does not rank as the as the largest problem facing humanity, even though the changes are likely to be large. In the short term, many other environmental concerns are already more worrisome, especially major ecological changes. Over the longer term, humanity's concerns will probably shift to the gradual depletion of irreplaceable "gifts" that Earth has freely provided, including fossil fuels, groundwater, and topsoil.

Chapter Eighteen

GLOBAL-CHANGE SCIENCE AND POLITICS

THE FUTURE GLOBAL WARMING will be large, but will it be bad or good? In terms of its effects on people, it depends. It depends on such things as who you are, where you live, what you do for a living, your ethical and aesthetic values, and your financial and economic status. Because these considerations lead to different value judgments, this question has no single answer.

Most climate scientists, aware of the limits of scientific knowledge and wary of complex value judgments, attempt to balance the facts and form their own conclusions. But these scientists are not the people the public usually hears from in media coverage of global climate change, precisely because of their "on-the-one-hand, on-the-other-hand" way of trying to balance complex issues. The media tend to prefer clever, crisply phrased sound bytes.

The public hears mainly from people toward the extremes of the global-warming issue, people who function as spokespersons for interest groups. These spokespersons cite results from scientific research, but they do so in a highly selective way, omitting the caveats that are part of a full scientific assessment, failing to place the results they quote in a larger context, and ignoring contradictory information that would frame a larger view. It is not particularly difficult to extract isolated scientific findings and string them together so as to support only one side or the other of a complex issue. The two extreme advocacy groups on the global warming issue can be called "environmental" and "industrial," even though these labels are obviously simplistic.

On the environmental side, the spokespersons are often heads of environmental groups. Day-to-day fund raising requires raising public concerns about the environmental degradation caused by human activities, and many groups do this in a responsible way. As scientists became aware of the likelihood of future global warming during the 1980s, the environmental movement added it to their list of concerns. But some extremists oversimplified the complexities of the global-warming issue and put out alarmist half-truths that the public heard or read out of context.

Extremists on the industrial side of the issue use the same technique. While many industries are run by people who are open-minded about environmental problems, some are not. In response to alarmist statements from the environmental extreme, industry supporters have in the last decade launched a counteroffensive that often errs in the opposite direction. They portray Earth as resilient to the puny impacts of humans and reject the growing body of evidence that global warming has begun and will be large in size.

The media seeking to inform the public on this complex issue often have to turn to spokespersons on the extremes. To claim "balanced coverage" and avoid charges of bias, the media tend to frame the debate by giving voice to both extremes, but the accusations from the two extremes do not add up to a coherent view of the issue. The public is understandably confused.

I offer two examples of the kinds of complexities and imbalances common in these exchanges. The first example is the sea-level rise that is occurring now and will occur in the future as mountain glaciers and other ice melts, and as warming ocean waters expand. A sea-level rise of about one-half meter is likely to occur within the next century, and perhaps a meter or more in the century after that. Advocates toward the environmental extreme, with help from the media, at times exaggerate this threat. In recent years, single blocks of ice equal in area to Rhode Island have broken away from the ice shelves on the margins of Antarctica. The media quote environmental spokespersons who interpret these events as signaling the impending destruction of the Antarctic ice sheet. Accompanying maps show that most of Florida would be underwater if all of the ice on Antarctica melted.

The truth is much less exciting. For one thing, those large ice shelves float in the ocean, and anything that floats in the ocean has already displaced water and raised the level of the ocean. So whether the ice shelf is still attached to Antarctica or has broken away to melt in warmer ocean waters to the north is irrelevant to sea level. The only way to make sea level rise is to melt the ice stored back up on the continent.

The fact that a huge chunk of ice has broken off also sounds ominous but is not. The ice shelves around Antarctica are continually being replenished by ice flowing outward from the interior of the continent, and that ice in turn is continually being replenished by snowfall. In other words, ice is moving through the system all the time without necessarily causing any change in the size of the ice sheet. So the fact that huge chunks of ice break off from time to time is a normal part of a system that in a long-term sense is stable. Only if most of the ice shelves around Antarctic were found to be steadily disintegrating in a way that had not occurred in previous centuries would we have reason to infer a trend driven by a greenhouse warming caused by humans. From what we know now, such a trend is not occurring. For now, the ice on Antarctica appears to be stable.

Still, with sea level now slowly rising, environmental spokespersons warn about its effect on highly populated coastal cities like Miami Beach, where sea walls already protect high-rise structures along the ocean. As global sea level continues to rise, they predict that disasters will be inevitable when new hurricanes strike. In response, global-warming skeptics acknowledge that future hurricanes may bring disasters, but they point out that sea-level rise will not be the main culprit. The next large hurricane to hit a highly developed area will bring a storm surge of perhaps 5–6 meters (15–20 feet), along with strong and destructive winds. This entirely natural phenomenon will cause far more destruction than the few tens of centimeters of sea-level rise caused by global warming. The real problem will be that far too many people have been allowed to build too close to the coast. Most of the disaster will result from the convergence of a natural phenomenon and ill-advised building codes.

The sea-level rise caused by global warming is a more serious concern for some less-affluent countries. For the small populations now living on low-lying coral atolls in the Pacific Ocean, and for the much larger populations living just above sea level in Bangladesh and other Asian river deltas, the greatest disasters of the future will again be those of natural origin (typhoons) as they impact vulnerable areas where populations have increased. But in these cases, the modest rise in sea level caused by global warming will also impact these areas in a direct way. In both regions, a half-meter rise of sea level will flood a large amount of the available living area, forcing people to move.

A second example of the underlying complexity of the global-warming debate is circum-Arctic climate change. Large changes have occurred in recent decades in the Arctic, including the gradual retreat and thinning of sea ice and seasonal snow cover, and the melting of permafrost (chapter 16). If this trend continues as climate warms, the Arctic will be transformed in major ways.

The implicit industry view is that the polar regions are a small part of Earth's surface and that few people live there. This view also emphasizes the likely beneficial impacts from polar warming, including fewer severely cold air masses in winter, and consequently fewer outbreaks of polar air into middle latitudes, thus permitting longer growing seasons in regions like Alaska and Siberia. Another benefit is the likely opening of Arctic ports and trade routes as the sea ice retreats.

Environmental advocates point out that future warming in the Arctic is projected to be much larger than for the planet as a whole, and that people who live in these environments will be heavily impacted. Ecosystems adapted to the retreating sea ice will be vulnerable. In Arctic lands, the permafrost underfoot will increasingly be half-melted slush and mush.

Both viewpoints have validity, and weighing all of the pros and cons is difficult. The main problem is that each extreme fails to mention the other side of the issue in pushing its own agenda. In my opinion, the harm we may do to Arctic ecosystems (and the people who depend on them) seems of greater importance than the economic benefits from growing extra food in midlatitude regions that for the most part already have large food surpluses. Arctic ecosystems have had over 3 million years to adjust to the presence of sea ice along the coastlines in the winter. When and if sea ice no longer reaches the coastlines in winter, and when and if it largely disappears from most of the central Arctic in summer, disruptions to these ecosystems (and the cultures that depend on them) are likely to be severe. Denial on the part of industry extremists is not a sufficient response.

For the public at large, opinions about global warming will likely reflect the value systems that individual people bring to this issue. Which of the possible concerns looms as more important—the impact of future changes on people and their environments, or the cost of preemptive actions to counter these impacts? For two very different views on global warming, read *Laboratory Earth* by Steve Schneider and *The Satanic Gases* by Pat Michaels and Robert Balling.

The global-warming debate came to a head with the 1997 Kyoto Treaty, which committed the industrialized nations of the world to reduce their rising CO_2 emissions to 1990 levels by the year 2010. The proposed reductions would have come mainly by cutting CO_2 emissions from coal burned in factories and power plants, by reducing gasoline consumption in cars and trucks, and by reducing consumption of oil and natural gas in homes. Among the few nations refusing to ratify this treaty was the United States, the largest single source of CO_2 emissions in the world. This failure enraged environmentalists, who saw it as a sign of selfishness, arrogance, and insensitivity to the environment and to the people whose lives would be changed by global warming.

The probusiness side countered that the reductions in CO_2 emission would lead to higher prices for home-heating fuel, gasoline for cars, and other day-to-day expenses but would have a very small effect on Earth's future temperature. The projected reduction in CO_2 (relative to the otherwise-projected increase) would avoid (or delay) a warming of about 0.1° to 0.2°C late in the current century. That amount is only 5 to 10 percent of the increase in global temperature predicted to occur by that time. It is also just a small fraction of the 0.7°C increase in global temperature during the last century and a quarter. Another problem with the treaty was that several less-industrialized nations were exempted from the Kyoto requirements, including China, then just beginning to emerge as an industrial power.

As a result, the U.S. Senate voted 95–0 in favor of a motion against signing any treaty that would commit the United States to environmental actions unless all the nations of the world shared in the burden. Politicians generally vote for policies their constituents want. Since most polls suggest that Americans favor action to protect the environment from the threat of global warming, why has the government avoided signing the Kyoto Treaty?

One explanation is that the benefit from the Kyoto reductions would occur decades in the future and even then would be undetectably small in people's lives, but the costs would be felt right away in higher prices in our daily lives. Politicians tend to favor actions that work the other way: benefits that are immediate and costs that lie far off in the future. A second plausible explanation is that politicians realize that the public support for proactive environmental policies is broad but not necessarily deep. Many of us favor doing something about global warming as long as the degree of sacrifice is acceptably small. But if sacrifice means trading in a big SUV for a small economy car, taking public transportation, or turning our thermostats down a little in winter or up a bit in summer, support for proactive policies begins to fade in the polls.

I think that this same line of reasoning leads to an unspoken truth about global warming that for some reason politicians of both parties ignore. To reduce current and future greenhouse-gas emissions to levels that would avoid *most* of the projected future warming, draconian economic sacrifices would have to be enacted that almost everyone would find intolerable: much more expensive fuel for travel and heating, much lower/higher thermostat settings in houses and workplaces, and extremely costly upgrades (or total replacements) of power plants. The drag on the economy and on quality of life from such efforts would be enormous, and few citizens would stand for it. At this time, with current technologies, we simply cannot afford the effort that would be required to mitigate the main impact of global warming. While not an excuse for doing nothing, this underlying reality should put the current debate in a clearer perspective.

It is even fair to ask whether most people would really prefer to avoid global warming. Most people complain about the onset of winter but greet the start of summer. Those living in snow-free sun-belt areas feel relief watching news coverage of snow and ice storms afflicting the Midwest or New England. Millions of people move south when they retire; few go north. If residents of our colder states were offered a simple up-or-down choice at the voting booth between (1) raising taxes to keep climate as cold as it is today, and (2) keeping taxes at current levels but allowing future Marchs to become more like Aprils and Novembers more like Octobers, which way would they vote? I doubt it would be to pay more to keep it colder. Any large-scale program to attack global warming would eventually run into underlying attitudes like these.

I think the single most effective thing that can be done about global warming is to invest in technologies that will reduce carbon emissions, especially those that will come from the 200-year supply of coal we will eventually burn. As standards of living in regions like Southeast Asia rise toward Western levels, all of the available fossil carbon will be burned within just a few centuries. The clearest way to maintain socially acceptable standards of living but still permit this development will come from technological breakthroughs that allow us to use fossil fuels but emit less CO_2 while doing so. Although I am optimistic that we will discover new technologies, I am pessimistic that they will be able to dispose of the billions of tons of CO_2 in a cost-efficient way. Investments in alternative energy sources also make good sense, but such sources seem likely only to delay the burning of our fossil fuels by a few decades, rather than replace them entirely.

This issue is both complex and unpredictable. Perhaps delaying some of the future warming will give us more time to find a technological fix. And if the warming arrives more slowly, maybe it will not reach quite as high a peak a few centuries from now. But just now, real solutions seem to be either expensive or just optimistic dreams for the future.

In the meantime, in the arena of public debate about global change, advocates on both sides of the issue are doing what they so loudly do. Aldo Leopold wrote in *A Sand County Almanac* that "Every profession keeps a small herd of epithets and needs a pasture where they may run at large." For the global-warming issue, those epithets are now running wild in the public domain.

Environmental extremists claim that the hands of industry spokespersons are soiled by financial support from coal utilities and oil companies driven by greed. Yet they refuse to acknowledge that the more aggressive actions implicit in their own position would have major (and probably unacceptable) costs to the public. Advocates for the environment often frame their positions with high-minded, preachy appeals to Jean Jacques Rousseau's notion of the "noble savage," the concept of a primitive but wise people who once lived lightly on the land and in complete harmony with the environment. They contrast this supposedly once-pristine world with the evils of heavy industrial development during the last two centuries. They portray industrial development as the first, and only, real human assault on nature.

This book has shown that the concept of a pristine natural world is a myth: preindustrial cultures had long had a major impact on the environment. First came the improved hunting skills that drove most large mammals and marsupials to extinction on several continents (chapter 6). A few millennia later came the pervasive impacts that grew directly out of the discovery of agriculture. Land-use changes resulting from widespread deforestation and irrigation eroded and degraded soils (chapter 7). Agricultural practices millennia ago led to large releases of greenhouse gases (chapters 8 and 9) and changes in global climate (chapter 10). Early technologies were relatively primitive, and human populations numbered in the hundreds of millions rather than in the billions, but our preindustrial ancestors had a large environmental and climatic impact on this planet.

Indeed, a good case can be made that people in the Iron Age and even the late Stone Age had a much greater per-capita impact on Earth's landscape than the average modern-day person. Most people born 2,000 years ago had no choice but to make a living by farming, and farming in most regions meant clearing forests.

Estimates are that the average person cleared several dozen acres of wooded land over a lifetime. How many people reading this book have been personally responsible for clearing dozens of acres of forest?

I have, of course, simplified matters to make a point. Our modern use of fossil fuels represents a large addition to our individual (per-capita) carbon emissions. The real problem lies in our aggregate impact—so many people now live on Earth that the total emissions from all of our billions of "capitas" have grown much larger than the total impact of all of the Iron Age people.

On the other side of the debate, many industry extremists seem convinced that nature is so inherently capable of self-healing that it can take care of any supposedly "trivial" wound we inflict on it. This view ignores a vast body of evidence showing that humans have become a major force in altering Earth's environment at a large scale. Human impacts began millennia ago, grew sizable long before the industrial era, and have since become the largest single environmental and climatic force on Earth. Nature occasionally reminds us of its power with a drought, a flood, a freeze, or a heat wave, but year after year, human activities remain the larger force.

Until the past year or two, I kept a wary eye on both sides of the global-warming debate. I discredited the disinformation coming from both extremes of the issue and tried to weigh the solid evidence and form my own opinions. Very recently, however, I have become aware that this dispassionate detachment may be too idealistic. The debate has taken a surprisingly ugly turn.

My introduction to this problem was accidental. After my hypothesis was first published, science journalists asked whether my results were relevant to the policy debate over global warming. I told them that the global-warming issue was a hornet's nest, and I didn't intend to stick my hand into such a nasty mess. I also said that I was willing to predict how people at or near the two extremes of the global-warming issue would probably react. I said the industry side might claim that my results showed that "greenhouse gases are our friend" because they had apparently stopped a glaciation. And I said that the environmental side could counter that if a relatively small population of farmers had managed to produce a greenhouse-gas increase large enough to stop a glaciation, then where are we headed in the future as the much faster greenhouse-gas increase carries gas concentrations well outside their natural range of variation? Within months, both of these predictions came true: reports on my hypothesis appeared in both industrial and environmental newsletters, each making use of it for their own ends.

Because of the wide coverage of the hypothesis, my name had somehow been added as a recipient of several newsletters that take skeptical or contrarian (in effect, proindustry) positions on global change. These newsletters opened a window on a different side of science, a parallel universe of which I had been only partly aware. The content of these newsletters purports to be scientific but actually has more in common with hardball politics.

One technique is instant commentaries against any new scientific results that appear to bolster the case for global warming. Within days of publication of peer-reviewed scientific articles, opinion pieces appear debunking these contributions, in some cases impugning the objectivity of the (well-respected) journals in which they are published. The authors of these opinion pieces are often well-known climate-science contrarians or others in related fields such as economics. Most of these articles come from contrarian web sites that receive large amounts of financial support from industry sources. In many cases, the authors are paid directly by industry for the articles they write. Even though many of these attacks amount to pin pricks that leave the basic conclusions of the criticized paper intact, the commentaries state or imply that the original results have been completely invalidated. In politics, this kind of counteroffensive is called "oppo research." A related technique is to cite published papers that address the same subject but come to conclusions more favorable to the industry view. In the cases where I know the science reasonably well, these papers do not match the rigor of the originals.

This alternative universe is really quite amazing. In it, you can "learn" that CO_2 does not cause any climatic warming at all. You can find out that the world has not become warmer in the last century, or that any warming that has occurred results from the Sun having grown stronger, and not from rising levels of greenhouse gases. One way or another, most of the basic findings of mainstream science are rejected or ignored.

In my opinion, some climate contrarians once served a useful purpose by pointing out alarmist exaggerations from environmental extremists. But this alternative universe is new and worrisome; in the name of uncovering the truth, it delivers an endless stream of one-sided propaganda.

Why would scientists devote time and energy to doing this kind of thing? One obvious possibility is money. Some industries pay scientists to write opinionated commentaries and give opinionated talks. Some environmental groups do the same thing. Financial support does not in and of itself prove bias, but it hardly suggests intellectual independence, especially in the case of individuals who earn large portions of their income from such sources.

Another potential motivation is thwarted ego. Some spokespersons are scientists whose reputations in the scientific mainstream never amounted to much, or whose early career successes faded away. Disappointed by the lack of recognition, they may have chosen to make a new "mark" by taking a different, far more publicly visible, path. Resentment over lack of mainstream success may also help to explain why these commentaries so often have a strident tone that mocks those with different views in a way that has no resemblance to the style of legitimate science.

Still another motivation may be the "white knight" or "hero" syndrome—the conviction that only heroic action in uncovering the "real truth" will save humanity from oncoming disaster or folly. Many contrarians appear to see mainstream scientists as dull-witted sheep following piles of federal grant money doled out by obliging federal program managers. In this view, only those who toe the party line that the global-warming problem is real, large, and threatening will get their hands on federal money. And of course only the lone visionary with clear vision can save the day.

This picture is a gross misrepresentation of science and scientists. Scientists are generally independent-minded individualists who think for themselves and instinctively resist the herd mentality. We do research for several reasons: mostly out of simple enjoyment, but also in the hope of being the first to discover something new and important. Much of our day-to-day progress comes in small incremental steps that produce new information used to test whether current ideas are valid or should be rejected. Occasionally we make breakthroughs by coming up with new ideas that replace old ones. In rare cases, a contribution may be so large and original that our names go down in the history of our field of science as a form of scientific immortality.

As a group, we are also notoriously unobsessed with personal wealth. I have often heard fellow scientists express delight that they are paid to do something they love so much, but no one I know makes a personal fortune competing for federal money to do basic research on climate. Many of us make a decent living, but we earn far less than we might in other careers. At most universities, even the top scientists earn salaries that are tiny fractions of the amounts paid to football coaches.

Scientists face stiff competition in obtaining grant money to support students and staff, to run labs and field programs, and for summer salary. At the National Science Foundation, success rates on proposals average less than one in five. I compare the competitive process of basic research to the challenges faced by people who run small businesses: we have to be better than the competition to succeed. This tough winnowing process leaves no room for the dull-witted sheep portrayed by the contrarian view.

Fortunately, most of the respected media outlets in this country seem to be ignoring the tidal wave of disinformation coming from the extremes of the global-warming argument. Apparently, the legitimate media recognize that these groups are lobbying on one side of a complex issue. Still, in a world where more and more people, especially the younger generation, get most of their news from the web,

I am concerned about the stealth impact of this lobbying effort. People exposed to an endless drumbeat coming only from one direction of an issue can be influenced.

The obvious question is whether anything can be done to provide more balance in this debate. Both First Amendment considerations and the protections of academic freedom argue against any top-down attempt to regulate such views, no matter how biased or one-sided. One recourse is to challenge the authors of these opinionated commentaries to be open with their audience. Have they been paid by an industrial or environmental group to write this particular article? More generally, what fraction of their grant income or salary over preceding years has come from industrial or environmental sources, as opposed to money won competitively from the government? This suggestion will likely be met with scorn by those who earn large portions of their salaries from interest groups, but at least their refusal to respond will serve as another warning sign of bias.

A big problem with this approach is that interest-group money is used to create or support innocuous-sounding citizen-action committees, with names along the line of "Geezers Concerned about the Environment." With these intermediary groups as cover, scientists can deny that they receive money from interest groups, even when interest-group money actually is the ultimate source. One way to uncover such evasions is to use the Freedom-of-Information Act to trace the flow of such money into "front groups." Recently, groups on the environmental side have begun doing just that, and the amount of money from industry sources appears to be in the millions of dollars every year, all with the apparent purpose of influencing the outcome of a supposedly scientific debate.

Probes by environmental groups into industry behavior are hardly an example of objectivity. Far preferable would be an open-ended investigation by well-respected journalists or other media into both industry and environmental meddling in this debate. My impression is that industry spends by far the largest amount of money for this purpose, but an aggressive and objective investigation into the facts would reveal the facts, and perhaps even shame the offenders.

The only other recourse available to those confused by the flood of opinions and counteropinions on this issue is simple common sense. If a person who is advocating a particular position is someone who always argues the same side of the issue and never balances the complex pros and cons, you can view what that person has to say with a very healthy dose of skepticism. And if a person repeatedly communicates these opinions through newsletters and web sites funded by industry or environmental interest groups, then again you can be skeptical. But if a person has a dual history of taking a persistently one-sided view and simultaneously receiving large amounts of income from only one of the interest-group extremes, then skepticism is too mild an attitude. People who fit this profile may or may not have been trained as scientists, but they are effectively functioning as lobbyists (paid propagandists) for one side of the global-change argument.

As should be clear from my personal experience (see chapter 11), I view challenges to new findings and new ideas as a normal part of the process of science. Still, I know of no precedent in science for the kind of day-to-day onslaughts and perversions of basic science now occurring in newsletters and web sites from interest groups. These attacks have more in common with the seamier aspects of politics than with the normal methods of science. Both the environmental and (especially) the industry extremists should leave the scientific process alone.

Chapter Nineteen

CONSUMING EARTH'S GIFTS

EVEN THOUGH I HAVE made the case that future climate change is likely to be large (chapter 16), I do not rank the oncoming global warming as the greatest environmental problem of our time. Other environmental issues seem to me far more immediate and pressing, and in the future I suspect our concerns will focus heavily on the eventual depletion of key resources.

One theme of this book is that humankind has been steadily transforming Earth's surface for some 8,000 years, initially in Eurasia and later on all continents. Initially, we caused these transformations by clearing land for farming; later, other aspects of civilized life joined farming as important causes of this transformation. Well before the industrial era, the cumulative result over many millennia was an enormous loss of what had been "natural" on this planet.

During the 1800s and 1900s, human population increased from 1 billion to 6 billion, an explosion unprecedented in human history. This rise came about because new sanitary standards and medicines reduced the incidence of disease and because human ingenuity led to innovations in agriculture that fed ever-larger numbers of people. As a result, our already sizable impact on Earth's surface increased at a much faster rate. Early in this millennium, we live in a world that has largely been transformed by humans.

By most estimates, the explosive population increase still under way will end near AD 2050 as global population levels out at some 9–10 billion people, or roughly 50 percent more people than now. A major reason for the predicted stabilization will be the increase in affluence that has historically resulted in fewer children per family. The good news is that we may avoid the worst of the predictions made by Thomas Malthus early in the nineteenth century about population-driven catastrophes like mass starvation and disease.

On the other hand, as affluence and technology continue to spread, increased pressures on the environment will occur for that reason alone. If a billion or more people in China and India begin to live the way Americans and Europeans do now, their additional use of Earth's resources will be enormous. Even without population increases, humanity will continue to alter the environment in new ways.

The cumulative impact of so many millennia of transforming Earth's surface has inevitably come at a cost to nature. Many of the problems our actions have created have been described elsewhere by ecologists and others knowledgeable about the environment. In the process of transforming Earth's surface, we have fragmented the space that ecosystems require, transported species from the places they belong to regions where their presence is invasive to existing flora and fauna, and caused the extinction of species in numbers that no one really knows.

Those who view the current or future impacts of human activities mainly through an economic prism generally focus on the benefits of taking and using what nature provides. This attitude is hardly new: we have been cutting forests and rerouting rivers to meet our basic needs for millennia. In most cases, our immediate needs have triumphed over any concern about what we might be doing to the environment.

Those with environmental concerns have in recent decades added a new argument to their side of this battle. Ecologists have coined the term "ecosystem services" to describe processes that nature provides for free and that have real economic value. For example, trees and other vegetation on hill slopes trap rainfall that would otherwise flow away and erode soils. As the retained water passes into subsurface layers, the soils slowly filter it and transform it into clean, drinkable water that can be retrieved from wells or springs. Some of the water flows into wetlands and is further cleansed there. Nature gives us a large supply of clean water.

But when the trees are cut or the wetlands are filled in, these free services are lost, and society must pick up the cost of doing nature's job. The extra runoff generated by removing trees requires construction of water-impoundment areas, which slowly accumulate silt and require maintenance. The loss of nature's subsurface cleansing of water requires municipal water treatment plants and home water filters, but these remedies rarely return water of the quality nature once provided. So we buy bottled water shipped from other regions or even other continents. All in all, we pay an economic price to replace ecosystem services. Ecologists rightly argue that the costs incurred from losing ecosystem services must be included in complete "economic" analyses of land-use decisions.

To this point, the costs of losing ecosystem services have not been overwhelming beyond relatively small regional scales. Thousands of years ago, irrigation-based cultures in the Tigris-Euphrates Valley were ruined by the buildup of natural salts carried by irrigation water. In arid environments, sparse rainfall cannot flush these salts from soils, which at some point become too saline for most plants to grow. Irrigation began in the Tigris-Euphrates region, and it also led to the first regional-scale abandonment of farmland. In AD 900–1200, the Mayans abandoned farmland in the Yucatan Peninsula because of a combination of droughts and cumulative nutrient depletion of the soils. Many other examples of exhaustion of resources at regional scales could be cited.

Resource exhaustion has to date been less obvious at a global level. Accessible deposits of some valuable minerals have been effectively depleted worldwide by mining, although most can still be recovered by more energy-intensive (deeper) mining at increasing economic and environmental cost. Many regions are now close to or beyond their natural limits of fresh water, especially clean drinking water, but again mainly at regional, not continental or global, scales.

Probably as a result of my long interest in Earth's climate history, my own concerns about the future tend to focus on a related set of longer-term problems— "gifts" that nature has provided us through slow-acting processes that took place well back in Earth's past and that cannot be replaced once they are consumed.

My concern about these gifts is simple: When these resources run low or run out, how will we find comparably inexpensive replacements? Like many people at retirement age, my personal concerns extend out to the life spans of my grandchildren. By the time they reach "old age," around the year 2075, I suspect it will be clear to all that some of the "free" gifts from Nature were not an unlimited resource. By then, I suspect that my grandchildren's generation will look back on the era between the late 1800s and the early part of the twenty-first century as a brief bubble of good fortune, a time when a lucky few human generations consumed most of these gifts, largely unaware of what they were doing.

We live today in an era of remarkably cheap oil, gas, and coal. It took nature hundreds of millions of years to create the world's supply of these resources, by burying organic carbon in swamps and inland seas and shallow coastal areas, and by cooking the carbon at just the right temperature and pressure. We only began using these resources in significant amounts in the middle 1800s, and yet the first signs are already at hand that the world will reach the year of peak oil production and consumption in just one or two decades, if not sooner (chapter 16). World supplies of natural gas will last a bit longer, and coal for a few centuries. I wonder whether we will ever find a substitute even remotely as inexpensive as these carbon gifts. We are investigating alternative sources of energy, but as of now none of them seems likely to be as widely available and as inexpensive as the solar energy stored in carbon-based fuels.

At some point early in the current century, gradual depletion of this vital commodity will presumably become a major economic issue. Once world oil production begins to decline by 1 percent or so per year, it seems likely to add a measurable cost to the functioning of the global economy, in effect adding a new form of built-in "inflation" on top of the normal kind. Fuel for cars and trucks is not the only concern; a vast array of products made from petrochemicals has become part of the basic fabric of our lives. All of them will cost more.

I also wonder about the long-term supply of water. With more than half of the supply of water from surface run-off already in use for irrigation and human consumption, we have for years been pumping water from aquifers deep in the ground, especially in arid and semi-arid regions into which many people are now moving. The water stored in the deep aquifers of the American West was put there tens to hundreds of thousands of years ago by melt water flowing southward from the margins of the great ice sheets, and by snow and rain water that fell during climatic conditions much wetter than today. In recent decades, the level of those aquifers has been falling as we extract water from depths where it cannot be quickly replenished by nature.

Pumping ever deeper will require more carbon fuel, which in turn will become more expensive. Gradually, the water from greater depths will contain ever-larger concentrations of dissolved salts that will be left on irrigated fields, making agriculture more difficult. Eventually, we will exhaust the useable water in these underground reservoirs. Little by little, agriculture will probably retreat from the arid high-plains regions of the West back toward the midcontinent regions nearer the Mississippi River, where natural rainfall supports agriculture. The same retreat will occur in other regions of extensive groundwater use across the globe.

I have no idea when these groundwater limits will be reached in each region, but news reports suggest the start of the problem may be close at hand in some regions. To conserve water, the municipal government of Santa Fe, New Mexico, recently began requiring home builders to retrofit six existing houses for improved water economy to offset the added water use in each new house built. In the Oklahoma-Texas panhandle region, oilman and investor T. Boone Pickens has been buying up the rights to groundwater in the glacial-age Ogallala aquifer. When municipal governments put extra burdens on local builders, and when wealthy oil tycoons invest in underground water as a scarce commodity, problems must be looming.

So I wonder if my grandchildren will live in a world some decades from now in which most of the economically accessible (and potable) groundwater has been extracted across much of the arid American west and other regions as well. I wonder if they will look back on this era as a brief bubble of relatively clean and incredibly inexpensive groundwater. Meanwhile, nature can't possibly replenish groundwater reservoirs fast enough to solve the problem. As with oil, once this "gift" is depleted in each region, it will effectively be gone forever.

I also wonder about topsoil. The most productive farms in the American Midwest can thank the ice sheets for their topsoil. Ice repeatedly gouged bedrock and scraped older soils in north-central Canada and pushed the eroded debris south, where streams of glacial melt water carried it into river valleys, and winds blew it across the western prairies. During the 1800s, farmers began breaking up the tough top layer of prairie sod, which had been held in place by extremely deep-rooted plants, and farming began at a scale the world had never seen. The midwestern American agricultural miracle has been one of the great success stories in human history.

But this great success came at a price. Repeated tilling exposed the rich prairie soils to decades of dry winds and floods. Estimates are that half of the original topsoil layer in the American Midwest has been lost, most of it flowing down the Mississippi River to the Gulf of Mexico. In the late 1900s, farmers began adapting new techniques that have reduced, but not stopped, the rapid removal of this precious gift. These methods and other conservation efforts will keep us from losing soil as rapidly as before, but slower rates of loss will continue to deplete soils that no longer have the natural protection provided by prairie vegetation.

Farmers and farm corporations now spend enormous sums of money every year on manufactured fertilizers to replenish nutrients lost to crop production and natural erosion. These fertilizers are produced from petrochemical (petroleum) products, which again brings us back to the gradual depletion (and increased expense) of carbon-based products in the coming decades. Once again, it will be a very long wait indeed until nature gets around to making more topsoil, probably 50,000 to 100,000 years until the next ice sheet bulldozes the next rich load southward.

I spoke out in the last chapter against environmental exaggeration and alarmism. With no wish to be alarmist about the slow depletion of these many gifts, especially carbon-based energy sources, I still wonder: What will humankind do when they grow scarce? Will our resourcefulness as a species open up new avenues? Or will the depletion be a true loss? I have no clue what answers to these vital questions the distant future will bring.

AFTERWORD TO THE PRINCETON SCIENCE
LIBRARY EDITION

FIVE YEARS HAVE PASSED since I wrote *Plows, Plagues, and Petroleum* (first published in 2005), and this new Princeton Science Library edition gives me an opportunity to look back on the way the science covered in the book has evolved. Because parts 1 and 2 provided fundamental background information, little has changed regarding the issues they discussed. Part 5 largely dealt with modern and future climate, and the most noteworthy shift in the last five years has been the development of an even stronger consensus that humans are the primary explanation for the approximately 0.7°C global warming during the last 125 years.

In contrast, the early anthropogenic hypothesis that was covered in parts 3 and 4 remains the subject of an ongoing debate that has generated several relevant new scientific findings. The dozen or more invited lectures I have given on this topic every year are one of several indications that interest in this issue is still at a high level. To date, the scientific community has not yet come to a consensus on whether the increases in carbon dioxide and methane during the last several thousand years were natural or anthropogenic in origin.

Here, I add a new summary updating the last half-decade of research on topics relevant to this continuing debate, with the new findings grouped into six major topics. Because some of these discussions are at times (unavoidably) a bit technical, they may be best suited for those interested in closely tracking the debate.

For the most part, these six topics track developments related to challenges to the anthropogenic hypothesis covered in chapter 11 of the book. By the time that chapter was written, just a year after the hypothesis first appeared, the major criticisms of the hypothesis had already been published. In the intervening five years, these criticisms have remained the same as before, and most of the critics seem to have moved on to other issues. As a result, the research summarized here explores new information or new thinking that responds to those early challenges. In several instances, these developments strengthen the case for the early anthropogenic hypothesis.

INSOLATION AND GREENHOUSE-GAS TRENDS

The early anthropogenic hypothesis had its origin in a simple observation: greenhouse-gas trends during this interglaciation (the Holocene) were different from those in the three preceding ones. During the first 10,000 years of interglacial stages 5, 7, and 9, the gas trends fell, but during the Holocene they fell for half or less of that interval before reversing direction and turning upward—CO_2 near 7,000 years ago and CH_4 near 5,000 years ago. Those reversals coincided with early farming activities that would be expected to have produced gas emissions: early deforestation that would have emitted CO_2, and early rice irrigation, livestock tending, and biomass burning that would have emitted methane. The centerpiece of the early anthropogenic hypothesis was the claim that early agriculture caused those greenhouse-gas increases.

Several studies soon countered the anthropogenic hypothesis with arguments keyed to interglacial stage 11, near 400,000 years ago. That interglaciation was chosen because its insolation trends were more similar to those of today than the three interglaciations I had selected. Because stage 11 orbital eccentricity values were as low as those in the Holocene, the amplitude of the eccentricity-modulated precession cycle at 22,000 years was more like the Holocene than those in interglacial stages 5, 7, and 9.

Several studies also concluded that gas emissions did not fall during the early part of stage 11, but remained high during the time that was most similar to the last few thousand years. These studies concluded that the warmth of the current interglaciation is likely to last another 16,000 years or so. In chapter 11 of the book, I responded that those studies had erred by neglecting to align stage 11 with the current (Holocene) interglaciation based on matching the insolation trends. Instead, they had aligned the early parts of the two preceding deglaciations and then counted forward in "elapsed time." When this method was tested against astronomically calculated insolation trends, it produced a major mismatch: the modern-day insolation minimum was aligned against a stage 11 insolation maximum. In contrast, when the two interglaciations were aligned based on insolation trends, I found that greenhouse-gas concentrations were falling during the interval in stage 11 that is most similar to recent millennia, consistent with the early anthropogenic hypothesis.

Now, several years later, ice-core drilling by the European Project for Ice Coring in Antarctica (EPICA) has recovered records that extend 800,000 years into the past. The fact that these records penetrate several additional interglaciations gives us several more chances to look at early interglacial greenhouse-gas trends. Two of these early interglaciations qualify for this comparison because they were preceded by rapid deglaciations that ended in a sharply defined early interglacial peak.

Methane and CO_2 trends during the six previous interglaciations are compared to those during the Holocene (stage 1) in figure A.1. The trends are plotted

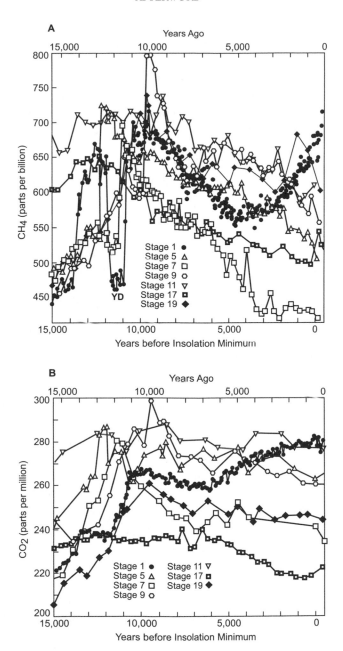

A.1. Trends of methane (A) and carbon dioxide (B) based on analyses of records from Dome C by the European ice-coring consortium EPICA, aligned according to their EDC3 time scale. Filled circles show trends during the current interglaciation. Other symbols show trends during the early parts of six previous interglaciations.

at a time scale compiled by the EPICA group. The first insolation minimum in each previous interglaciation is used as the "zero point" to align the records with the modern-day insolation minimum.

The CO_2 and CH_4 trends share several features. In the earlier parts of the records, the gas concentrations rose as the preceding glacial climates gave way to interglacial conditions. The concentrations then reached maximum values some 11,000 years or so prior to the "zero-point" alignment, at the start of the inter-glacial climate regimes. Following this peak, values then began to fall, some quickly, some slowly. For all six previous interglaciations, the gas concentrations continued to drop through the times equivalent to today. But for the Holocene, and only for the Holocene, the trends reversed direction and rose steadily for the last several thousand years. The Holocene CO_2 and CH_4 trends are thus unique compared to the previous interglaciations.

These trends pose a major problem for those who favor a natural explanation for the Holocene CH_4 and CO_2 increases. If the late-Holocene gas increases were caused by the natural operation of the climate system, similar trends would be ex-pected to have occurred early in the previous interglaciations, when the insolation forcing was similar. Yet not a single previous interglaciation in figure A.1 shows a CO_2 or CH_4 rise during the comparable interval.

Interglacial stage 19 is of particular interest because it is a closer insolation ana-log to the Holocene than any previous interglaciation, including stage 11. Stage 19 has the same low-amplitude precession signal as stage 1, but also a more simi-lar phasing of the separate insolation contributions from the tilt and precession cycles. In contrast, the tilt and precession cycles in stage 11 have a very different alignment from those in the Holocene. The CO_2 trends for stage 19 and the Holocene are very similar, starting at nearly identical glacial values of about 185 parts per million (not shown in figure A.1b), rising to early interglacial peaks in the 260–270 parts per million range, and then beginning similar decreases. As noted earlier, the Holocene trend reversed direction nearly 7,000 years ago and rose to a peak of 280–285 parts per million in late preindustrial time. In contrast, the stage 19 trend continued to fall and reached a value of about 245 parts per million at the time equivalent to the present day. This stage 19 value lies at the top end of the proposed natural range of 240–245 parts per million in the early anthropogenic hypothesis and is 35–40 parts per million lower than the approxi-mately 283 parts per million peak concentration reached in the latest Holocene (around AD 1200).

One mechanism by which science operates is falsification. Hypotheses cannot be proved, but they can be disproved by strong evidence. In this case, we have the results from six tests run by the climate system on the group of hypotheses that invoke natural causes of the Holocene greenhouse-gas increases. Because no comparable increases in gas concentrations occur during any of the six previous interglaciations, the natural hypothesis fails twelve successive tests—six for CO_2 and six for methane. Based on these failures, a strong case can be made that natural explanations (of any kind) have been falsified.

CLIMATE/CARBON MODELS

Models that attempt to simulate the interactions between the physical parts of the climate system and the major carbon reservoirs have been used to explore the cause of the Holocene CO_2 increase. These simulations, which have started from the assumption that the CO_2 increase was natural in origin, have simulated a late-Holocene CO_2 increase, but often for different reasons, including a decrease in (loss of) terrestrial biomass, an increase in ocean temperatures, construction of coral reefs, and delayed adjustments of ocean chemistry to imbalances imposed in earlier millennia.

All of these climate/carbon simulations face a difficult challenge that, to date, has not been overcome. The problem is that the models need to simulate not just the rising CO_2 trend during the Holocene but also the falling CO_2 concentrations during previous interglaciations (fig. A.1b). The challenge this poses, summarized schematically in figure A.2, is linked to the assumption used in these models that insolation forcing is the primary driver of CO_2 changes in all cases. If the relatively low-amplitude insolation trends during the Holocene drove the observed CO_2 increase, then it would seem only logical that the higher-amplitude insolation trends during previous interglaciations would have driven larger CO_2 increases at those times. But any simulated CO_2 increases (large or small) would be contrary to the trends found in ice cores.

At this point, only one attempt (by Guy Schurgers and colleagues) has been made to simulate CO_2 trends during both the Holocene and any previous interglaciation. This climate/carbon model simulated a 7 parts per million CO_2 increase during the Holocene (about a third of the observed amount), but it also simulated a 22 parts per million increase during the isotopic stage 5 interglaciation. Yet the CO_2 signal early in stage 5 shows no increase but instead a slight downward trend. This modeling attempt failed to meet the challenge.

Other such attempts to simulate previous interglaciations are underway as this is being written. I find it difficult to see how any model that simulates a natural upward CO_2 trend during the Holocene can avoid simulating an even larger upward CO_2 trend early in the previous interglaciations, contrary to ice-core observations.

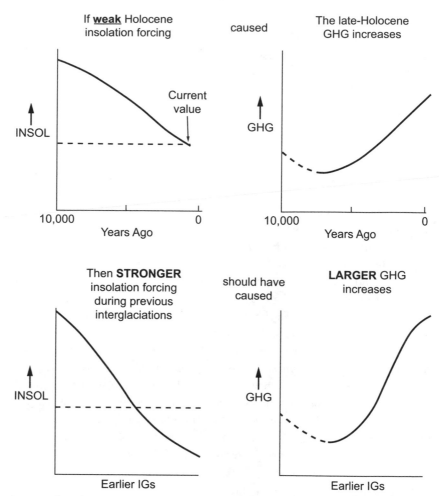

A.2. A thought experiment. If the small-amplitude insolation changes during the Holocene drove the observed CO_2 increase (top), then the larger-amplitude insolation changes during previous interglaciations should have driven larger CO_2 increases (bottom). Yet ice-core records show CO_2 decreases.

LAND USE AND POPULATION

Another general criticism of the early anthropogenic hypothesis has centered on the idea that there could not have been enough people alive so many thousands of years ago to have taken control of greenhouse-gas trends and driven them upward during subsequent millennia. For the start of the historical era 2,000 years ago, the best estimates point to a global population of about 200 million people.

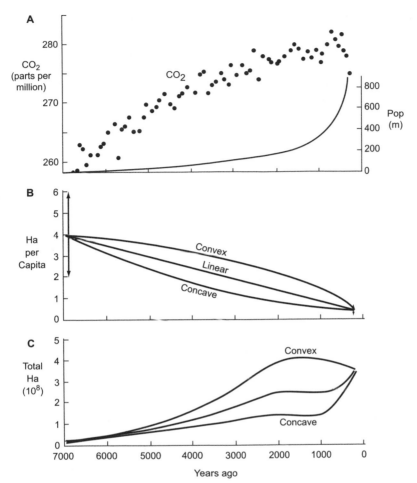

A.3. Trends in atmospheric CO_2 concentration and global population (A), estimated per-capita land use (B), and total (global) forest clearance (C) during the middle and late Holocene.

Prior populations are not known, but hindcasts based on an assumed population doubling time of 1,000 years suggest that perhaps 10 to 20 million people lived 6,000 years ago (fig. A.3a).

In chapter 7 (fig. 7.2) of the book, I showed that the beginning of the CO_2 increase (then dated to 8,000 years ago in Vostok ice, now at 7,000 years ago in Dome C ice) coincided with the first spread of agriculture and land clearance across the forests of Europe. But the early rise of the CO_2 signal that followed, and the leveling off of that early CO_2 increase after 2,000 years ago, bears little resemblance to the late-rising population curve (fig. A.3a). This mismatch is an obvious argument against the hypothesis that humans were the major cause of the CO_2 increase. As further support of this criticism, several model-based estimates of preindustrial land use that assume a close (i.e., nearly linear) relationship between population size and total land clearance have concluded that most global deforestation occurred within the last 300 to 400 years, during the time of the most rapid preindustrial population increase, and much later than the early CO_2 rise.

But this assumption that land use and population are linked in a close relationship is not supported by results from several field-oriented disciplines. Anthropological studies across a range of contemporary cultures that still practice forms of shifting cultivation, such as slash-and-burn farming, provide insights into farming practices used millennia ago, when all agriculture was likely of this form. Studies in land-use archeology, paleoecology, paleobotany, and sedimentology provide constraints on past changes in type and extent of agriculture, in gradual replacement of natural vegetation by domesticated crops, and in enhanced erosion of slopes bared by deforestation and tilling of fields. The common message from these field disciplines is that per-capita land use during the last 7,000 years has not remained constant (that is, tied to population in a linear way), but instead has decreased by a large amount.

Decades ago, Ester Boserup proposed a sequence summarizing how land use has changed with increasing population (table A.1). In the earliest and least populated phase of agricultural development (the long-fallow phase), early farmers set fire to patches of forest and planted seeds in ash-enriched soil between charred stumps. When soil nutrients became depleted after a few years, people simply moved on to another plot, and then another, returning to the original plot only after twenty to twenty-five years or longer. This kind of agriculture required little per-capita labor, but the continuing rotation among plots used a large amount of land.

Through time, as increases in local population densities reduced the amount of available land, farmers were forced to shorten the fallow period. At some point, with still-ongoing population increases, farming became restricted to the same plot of land every year (annual cropping). They began to make use of technological innovations that increased yields per acre, including improved plows, livestock traction, irrigation, and fertilizers. Ultimately, many farmers began rotating two or more crops per year in the same field and developing sophisticated and extensive irrigation systems. Despite the benefit of iron tools and other new technology, this later phase of intensive farming required very large amounts of labor per person—collecting and spreading manure and compost, tending the livestock that supplied most of the manure, eliminating weeds and insects, maintaining irrigation canals, and other such efforts.

TABLE A.1
The Boserup Sequence of Changing Holocene Land Use

	Shifting Agriculture		Intensive Agriculture	
Type	Long-fallow	Short-fallow	Annual cropping	Multicropping
Population	Low	>	>	High
Hectares/person	2–6	1–2	0.3–0.6	0.05–0.3

Figure A.3b summarizes one estimate of changes in per-capita land use along the "Boserup Sequence" proposed by myself and Erle Ellis, an expert on land use. During the early part of the sequence, when all those who farmed used the long-fallow method, per-capita land use was several hectares per person (1 hectare = 2.4 acres), with a best estimate of 4 hectares. At the later (preindustrial) end of the sequence, land use had fallen to an estimated 0.4 hectares per person, prior to reaching even lower average values (0.2–0.3 hectares) during the 1800s and 1900s. Because the intervening trend is not known, several possible trajectories are shown. We infer that the actual trend is likely to have some kind of convex shape, based on the abundance of evidence of widespread technological innovation during the historical era.

The total area of land cleared over time on a global basis can be estimated as the product of global population (fig. A.3a) and the trends in per-capita land use (fig. A.3b). All three trends (fig. A.3c) show global clearance leveling out nearly 2,000 years ago, and the convex case even shows a small reduction in total land use since 1,500 years ago. The most critical factor in the leveling out of these trends since 2,000 years ago is the decrease in per-capita land use, which cancels out part or all of the rapid rise in population. With allowance for the Boserup land-use trend through the late Holocene, the long-term trend in total land use (fig. A.3c) now looks more like the CO_2 signal (fig. A.3a).

Atmospheric CO_2 concentrations depend, however, on the *rate* of clearance rather than the *total amount*. In addition, other factors would have to be considered to make a full comparison between changes in land use and atmospheric CO_2 concentrations. Because early farming was concentrated in well-watered valleys with rich soils, forests cleared from those areas are likely to have been more carbon-dense than those on hillsides and slopes cleared in more recent times. Allowance for this gradual change in carbon density would further steepen the rate of the earlier Holocene CO_2 emissions compared to the more recent ones. Finally, realistic comparisons will have to take into account the very long residence time of CO_2 in the atmosphere. Some 15 to 20 percent of the total amount of CO_2 injected into the atmosphere remains there for countless millennia.

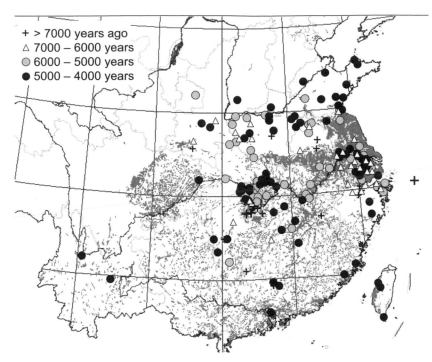

A.4. The spread of rice cultivation in China prior to 4,000 years ago based on several hundred archeological sites. The gray background pattern shows modern-day rice paddy areas.

A second area of doubt concerning the early anthropogenic hypothesis has been whether early agriculture can account for the reversal in trend of the methane signal nearly 5,000 years ago. In a joint effort with Zhengtang Guo and his colleagues in Beijing, I found new evidence, based on a compilation of several hundred archeological sites from China, that shows a large increase in the number of sites with rice remains at the time of the methane reversal (fig. A.4). Prior to 7,000 years ago, scattered sites contain remains of what are thought to be strains of dry-adapted natural rice that may have been replanted in slightly moister soils. Between 7,000 and 6,000 years ago, as rice remains begin to become somewhat more numerous in the archeological record, direct evidence of irrigation. such as radiocarbon-dated wooden sluiceways. first appears. Between 6,000 and 4,000 years ago, the technique of growing wet-adapted rice strains in irrigated paddy fields spread rapidly across the area in south-central China where irrigated rice is grown today.

The number of sites dating to the interval between 5,000 and 4,000 years ago exceeds those dating to between 8,000 and 7,000 years ago by a factor of 10, an increase rapid enough to suggest a causal link with the reversal of the methane trend nearly 5,000 years ago. In addition, archeologists have concluded that rapidly expanding rice farming in south-central China and dry-land agriculture in north-central China caused a major surge in population during these millennia. These population increases would have produced still more methane from additional factors such as the growing numbers of livestock tended, greater biomass burning, and more generation of human waste.

Although these archeological compilations support the idea that humans played a role in the methane reversal nearly 5,000 years ago, this link cannot easily be quantified. The number of individual rice paddies in China (and in all of southern Asia) today is incalculable—probably at least in the tens of millions—whereas the archeological database consists of just a few hundred ancient sites located near paddies whose actual sizes are, in most cases, unknown.

CARBON BUDGETS AND CO2 FEEDBACK

The most telling criticism of the early anthropogenic hypothesis has been the conclusion that even deforestation of much of southern Eurasia, along with substantial parts of the Americas and Africa, could not have driven CO_2 concentrations upward by about 20 parts per million or more prior to the industrial era (fig. A.1b), much less explain the full 35–40 parts per million size of the CO_2 anomaly I proposed. In chapter 11 of the book, I conceded that my critics (most notably Fortunat Joos and colleagues) were right. And in a paper published in 2007, I concluded that direct anthropogenic emissions from deforestation, and lesser contributions from early coal burning in China and severe degradation of soil profiles in parts of Eurasia, could not account for more than about 9 to 10 parts per million of the 35–40 parts per million CO_2 rise, or about 25 percent of the total.

Although some scientists inferred that this concession invalidated my hypothesis, they missed the fact that the CO_2 trend in the last 7,000 years still departed from those in previous interglaciations by the same 35–40 parts per million. As a result, the anomaly had not disappeared. These two observations presented a major enigma: a 35–40 parts per million anomaly that appears to have been anthropogenic, but a direct contribution from deforestation limited to only 9–10 parts per million.

In chapter 11 (fig. 11.4), I proposed that the answer must lie in feedbacks operating within the climate system that boosted the direct CO_2 contributions from deforestation and early coal burning by releasing additional CO_2. My reasoning was that the warming of the atmosphere caused by the *direct* anthropogenic emis-

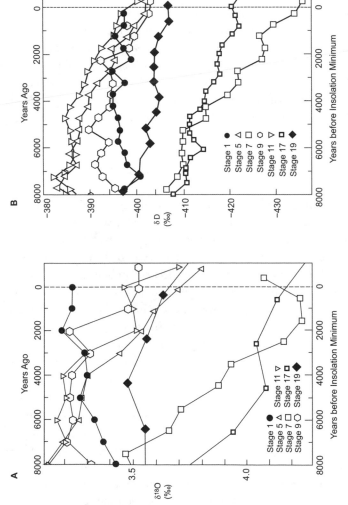

A.5. Evidence that the current interglaciation has remained warmer than the trend toward cooler climates during previous ones: (A) oxygen-isotopic measurements ($\delta^{18}O$) of deep-sea benthic foraminifera, with lighter (more negative) values indicating warmer temperatures; (B) deuterium (δD) measurement from Dome C ice cores, with lighter values indicating colder Antarctic air temperatures.

sions of greenhouse gases (both CO_2 and CH_4) would also have warmed the ocean, or at least have kept it from cooling. In this way, an anomalously warm ocean could then provide positive CO_2 feedback to the atmosphere via an indirect, but still anthropogenic, effect.

At first glance, this explanation may seem unlikely. How could a 9–10 parts per million initial CO_2 "push" from direct anthropogenic emissions produce another 26–30 parts per million of CO_2 feedbacks? But framing the problem this way leaves out methane, which also plays a prominent role. The estimated anthropogenic CH_4 anomaly from farming was 250 parts per billion, which would have had an effect on climate equivalent to roughly 12 parts per million of carbon dioxide. With this additional help from methane, direct greenhouse-gas emissions equivalent to approximately 21 to 22 parts per million of CO_2 would have given the climate-carbon system an initial push that resulted in the CO_2 feedbacks totaling 26 to 30 parts per million.

In chapter 11, I also speculated that the most likely source of CO_2 feedback is the ocean, particularly the deep ocean and the Southern Ocean, both of which are known to have played a significant role in natural CO_2 changes during glacial-interglacial cycles. New evidence now supports this suggestion.

Oxygen-isotope ($\delta^{18}O$) ratios from bottom-dwelling foraminifera trend toward heavier values during all six previous interglaciations (fig. A.5a). This trend indicates either that new ice sheets were growing or that deep water was cooling, or (more likely) both. In contrast, the $\delta^{18}O$ trend during the Holocene since 7,000 years ago has been toward lighter values. Because ice sheets were neither growing nor melting in significant amounts during this time, this trend toward lighter values suggests that deep-ocean temperatures warmed during the middle and late Holocene, before cooling slightly in late preindustrial time. The approximately $0.2°/_{oo}$ negative trend would translate to a deep-ocean warming of about $0.84°C$. Because CO_2 is less soluble in warmer water, an ocean warming would have released CO_2 to the atmosphere and driven concentrations higher.

Deuterium (δD) ratios from Antarctic ice cores show a similar "warm anomaly" during the Holocene. In the early parts of all six previous interglaciations, the ratios drifted toward more negative values, indicating that air temperatures directly above the Antarctic ice sheet were cooling (fig. A.5b). In the Holocene, however, the ratios varied only slightly and ended up at nearly the same value as 7,000 years ago. In this case, the Antarctic registered a "warm anomaly" in the sense that it failed to register the "normal" cooling observed in previous interglaciations.

As a result, ocean temperatures near Antarctica would also have remained warmer, and sea ice would have failed to advance as it did early in previous interglaciations. These changes would have kept Southern Ocean waters in greater

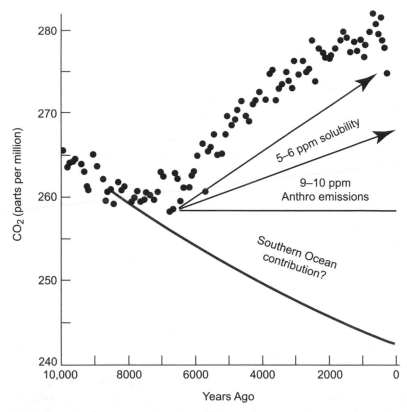

A.6. A "pie-chart" representation of possible contributions to the anomalous CO_2 trend during the last 7,000 years from: warming of the deep ocean (resulting in decreased CO_2 solubility), direct anthropogenic emissions, and maintenance of anomalous warmth in the Southern Ocean.

contact with the atmosphere, which, according to several conceptual models, would have kept CO_2 concentrations from falling. The size of this additional CO_2 feedback effect is not known.

New modeling results published in collaboration with atmospheric scientists John Kutzbach, Steve Vavrus, and Gwenaelle Phillipon at the University of Wisconsin support the evidence from marine and ice-core observations. We ran an experiment with an atmospheric general circulation model that included an interactive dynamical ocean in order to assess the effect of early anthropogenic greenhouse gases on the atmosphere and ocean, and the simulation produced warming effects similar to those from the oxygen-isotope and deuterium evidence. The early anthropogenic greenhouse gases would have warmed the deep ocean by an amount slightly smaller than that indicated by the oxygen-isotopic trends (fig. A.5b). They would also have made the atmosphere over Antarctica warmer by an amount that closely matched the anomalous warmth indicated by the deuterium data (fig. A.5a).

A deep ocean that warmed in the Holocene by 0.5–0.84°C would have released additional CO_2 into the atmosphere and raised concentrations by about 5 to 6 parts per million. This additional CO_2 release would fill much of the gap left between the direct anthropogenic emissions and the observed trend (fig. A.6), especially if the lower CO_2 value in latest preindustrial time (1600–1750) is used as the "target." At this point, it is not possible to quantify the carbon exchanges in an Antarctic region that stayed warmer in the Holocene compared to previous interglaciations. As a result, we do not yet know whether additional CO_2 from a warm Southern Ocean can fill the remaining "carbon gap" needed to explain the full 35–40 parts per million CO_2 anomaly (fig. A.6).

Is a New Glaciation Overdue?

The original anthropogenic hypothesis also included the claim that new ice sheets would have begun forming by now if CO_2 and CH_4 levels had fallen to their natural levels rather than being driven upward by emissions from early agricultural activities. In chapter 10 of the book, I reported results from an experiment run with the GENESIS general circulation model with coauthors Steve Vavrus and John Kutzbach at the University of Wisconsin. That experiment simulated year-round snow cover in a few grid boxes along the high spine of Baffin Island. We regard the occurrence of year-round snow cover that can thicken through time as evidence of glacial inception. I commented that I had been hoping for a home run with this simulation, but the outcome felt more like beating out an infield single.

In subsequent years, Steve Vavrus, John Kutzbach, and I, along with Gwenaelle Phillipon, have run several simulations with the National Center for Atmospheric Research's Community Climate Model (version CCSM3). In all experiments, the control-case conditions were altered by lowering the atmospheric CO_2 concentration to 240 parts per million and the CH_4 value to 450 parts per billion to match the values proposed in the early anthropogenic hypothesis. One simulation used the basic atmospheric version of the model, while others included vegetation/albedo feedback, ocean dynamics (noted above), and higher-resolution topography.

All of these experiments simulated year-round snow cover over various high-altitude and high-latitude regions in Canada and northern Eurasia. For example, the dynamical-ocean experiment simulated twelve-month snow cover across the

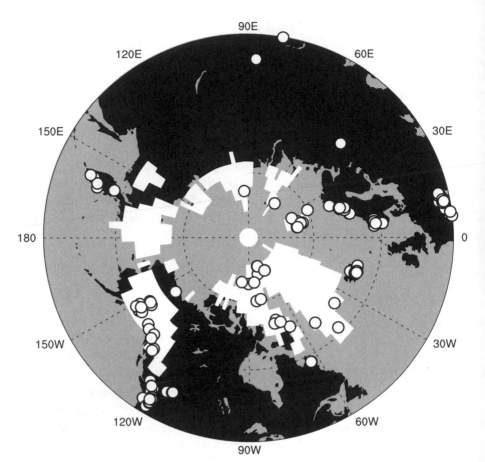

A.7. Regions with year-round snow cover (incipient glaciation shown in white) from a general circulation model simulation run with atmospheric CO_2 set to 240 parts per million and CH_4 to 450 parts per billion, the values that the early anthropogenic hypothesis predicts would occur now in the absence of human intervention. White circles are modern-day ice caps and mountain glaciers.

northern Rockies, the Canadian archipelago, Baffin Island in northeastern Canada, eastern Siberia, and several Arctic islands north of Siberia (fig A.7). The total area (outside Greenland) with year-round snow in that experiment was 30 percent larger than the size of the modern-day Greenland ice sheet.

Many of the snow-covered grid boxes surround locations of modern-day ice caps and mountain glaciers that are now melting back from more extensive preindustrial (Little Ice Age) limits. Such a configuration makes sense from comparing the relative levels of the greenhouse gases. If Little Ice Age glaciers and ice caps were able to form or grow across small regions when the CO_2 concentration was around 270–275 parts per million and the CH_4 level was 670–690 parts per billion, then much more extensive snow cover would be expected with CO_2 set to 240 parts per million and CH_4 to 450 parts per billion in the experiment.

The results from these experiments felt like at least an extra base hit, but it would be premature to start a full "high-five" celebration of a successful prediction. The CCSM3 model has a moderate cold bias in summer that tends to favor "extra" snow cover, and the control-case simulation for that model with "modern" greenhouse-gas levels places year-round snow cover on a few grid boxes on islands in the Eurasian Arctic, where snow actually disappears in summer today and older ice caps are melting. To be certain that our findings are not the result of model bias, it will be necessary to run similar experiments with other models, including the new CCSM4 model soon to be made available at NCAR.

PANDEMICS, MASS MORTALITY, AND CO2 OSCILLATIONS

In a corollary to the early anthropogenic hypothesis, I claimed in part 4 of the book that several dips in CO_2 that lasted for several centuries were not entirely natural in origin but resulted at least in part from well-documented episodes of mass mortality during the historical era. The central idea was that mass mortality allowed forests to reoccupy abandoned land, sequester carbon from the atmosphere, and drive CO_2 levels lower. At that time, only two CO_2 records from Antarctica were available to test this idea: a well-dated record from Law Dome that extended back slightly less than 1,000 years and a poorly dated record from Taylor Dome that extended back through the entire historical era and beyond. The Taylor Dome ice had not been analyzed for the presence of volcanogenic layers needed to improve the dating.

In 2006, MacFarling Meure and colleagues in New Zealand published a very detailed and well-dated CO_2 record from Law Dome (fig. A.8). The new analyses spanning the last 1,000 years fit seamlessly into the earlier ones, and the earlier part of the record was pushed back to nearly 2,000 years ago. Plotted below the CO_2 record is a summary of the major mass-mortality events of the historical era (prior to industrialization). The correlation between the two signals is striking: every major CO_2 decrease lines up with an episode of mass mortality in Europe and China or the Americas. Intervals of stable or increasing CO_2 are free of mortality disasters.

As noted in the book, pandemics are the major factor in these depopulation episodes: bubonic plague and other outbreaks in the late Roman era (200–600), bubonic plague again (the "Black Death") in late medieval times (1348–1400),

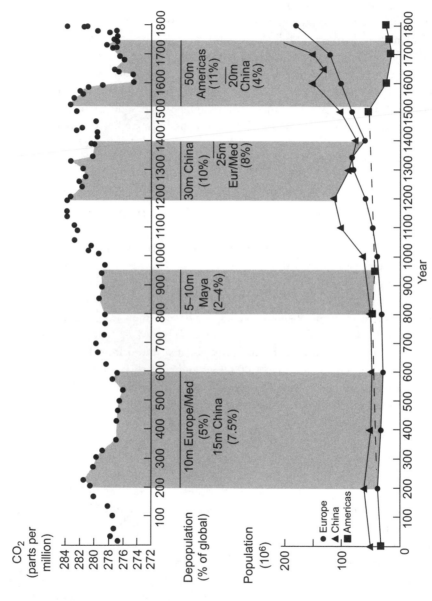

A.8. High-resolution CO_2 analyses from Law Dome Antarctica show CO_2 decreases that match episodes of mass human mortality caused by pandemics in Europe and the Americas and by civil strife in China.

and the "virgin soil" epidemics brought to the Americas by Europeans (1492–1700). Subsequent work with Ann Carmichael of Purdue University, an expert on the history of diseases, made it clearer to me that the two depopulation episodes in China were probably not the result of pandemics, but of extreme civil strife caused by invasions from Mongolia, including Genghis Kahn and his descendants in the 1200s.

Correlation is not causality, but the excellent match in figure A.8 certainly supports the proposed link between mass mortality and CO_2 decreases. In contrast, this more detailed CO_2 record raises questions about natural explanations for the CO_2 decreases, especially the drop of nearly 10 parts per million from high values in the medieval era around 1200 to the low values in the Little Ice Age between 1600 and 1700. As I noted in chapter 12 of the book, carbon-climate models suggest that each 1°C decrease in global temperature should cause a 12 parts per million drop in CO_2 because of increased CO_2 solubility in the cooler ocean and also in part from decreased oxidation and carbon release from vegetation litter on land. Proxy-based estimates of Northern Hemisphere temperatures now bracket the Northern Hemisphere cooling from 1200 to 1700 between 0.2° and 0.5 °C, with a mid-range estimate of 0.3–0.35 °C. According to the models, a cooling of that size could account for a CO_2 drop of 3 to 4 parts per million, but this estimate would leave the other two-thirds of the observed decrease to be explained by other factors, such as anthropogenic forcing.

My attempts in recent years to simulate the effects of sequestering billions of tons of carbon on abandoned farmland suggest that reforestation could directly explain at least 4 parts per million of the observed 10 parts per million drop. This analysis left open the possibility of a larger anthropogenic amount if CO_2 feedback from the ocean (especially the Southern Ocean) boosted the direct response.

Improving on these estimates would require better data on the per-capita "forest footprint" of the farmers who died in order to calculate the resulting carbon sequestration more accurately. Another uncertainty that needs further attention is the question of how much farmland that might have been abandoned when its farmers died was instead taken over by family members or neighbors and thus prevented from reverting to the wild. Population densities in the Mediterranean region during the late Roman era were probably high enough that substantial re-occupation of this kind occurred. By the time of the Black Death pandemic, much of west-central Europe was also populated at high enough levels that reoccupation would have been common, although not in less populated northeastern regions like the Baltic and Russia. This reoccupation, along with the amazingly rapid recovery to prepandemic population levels by 1500, might explain why the imprint of the European Black Death on the CO_2 signal is subtle.

In contrast, the 85 to 90 percent mortality rates of indigenous people in the

Americas between 1500 and 1700 left almost no one to reoccupy abandoned farmland, and the land would have reverted to forest. This pandemic likely played an important role in the CO_2 drop to low concentrations near 1600. Because European populations in the Americas remained low until 1750–1800, the areas reforested by this massive pandemic remained reforested for centuries.

In summary, the years since publication of *Plows, Plagues, and Petroleum* have seen numerous follow-up efforts to test and explore various aspects of the early anthropogenic hypothesis. In my (not entirely unbiased) opinion, these new contributions have had two results. First, the weaknesses of natural explanations of the Holocene greenhouse-gas increases have become clearer. The central weakness remains the continuing lack of an answer to a very simple question: If the upward trends in late-Holocene CO_2 and CH_4 concentrations were caused by natural factors, why do similar intervals during previous interglaciations fail to show upward trends? These multiple failures falsify the natural explanations.

Second, new research has addressed the most serious criticisms of the early anthropogenic hypothesis. Although these studies have not "clinched the case," they have provided evidence that points to plausible answers to all of the major criticisms. The net effect of this new research has been to keep the early anthropogenic hypothesis alive and well (and unfalsified).

BIBLIOGRAPHY

The following volumes were useful sources in compiling this book and are written in a general style that should be accessible to nonspecialist readers.

Part Two

Alvarez, W. *T. rex and the Crater of Doom.* Princeton: Princeton University Press, 1997.
Chorlton, W. *Ice Ages.* Alexandria, VA: Time-Life Books, 1983.
Imbrie, J., and K. Imbrie. *Ice Ages: Solving the Mystery.* Short Hills, NJ: Enslow, 1979.
Tudge, C. *The Time before History.* New York: Touchstone, 1997.

Part Three

Diamond, J. M. *Guns, Germs, and Steel.* New York: W. W. Norton, 1999.
Roberts, N. *The Holocene.* Oxford: Blackwell, 1998.
Smith, H. *The World's Religions.* San Francisco: Harper, 1991.
Williams, M. A. *Deforesting the Earth.* Chicago: University of Chicago Press, 2003.

Part Four

Bray, R. S. *Armies of Pestilence.* New York: Barnes & Noble, 1996.
Cartwright, F. E. *Disease and History.* New York: Dorset Press, 1972.
Denevan, W. M., ed. *The Native Population of the Americas in 1492.* Madison: University of Wisconsin Press, 1978.
Krech, S. *The Ecological Indian.* New York: W. W. Norton, 1999.
McEvedy, C., and R. Jones. *Atlas of World Population History.* New York: Penguin Books, 1978.
McNeil, W. *Plagues and Peoples.* Garden City, NY: Doubleday Press, 1976.
Times Atlas of World History. Maplewood, NJ: Hammond, 1982.

Part Five

Deffeyes, K. *Hubbert's Peak.* Princeton: Princeton University Press, 2001.

Epilogue

Leopold, A. *A Sand County Almanac.* New York: Oxford University Press, 1949.

Michaels, P. J., and R. C. Balling, Jr. *The Satanic Gases.* Washington, DC: Cato Institute, 2000.
Schneider, S. H. Laboratory Earth. New York: Basic Books, 1997.

AFTERWORD

Boserup, E., *The Conditions of Agricultural Growth.* New York: Aldine, 1965.

General resource for information on Earth's climatic history:

Ruddiman, W. F. *Earth's Climate.* New York: W. H. Freeman, 2001.

The following peer-reviewed publications were used as the basis for sections of the book.

PARTS THREE AND FOUR

Ruddiman, W. F. "Cold Climate during the Closest Stage 11 Analog to Recent Millennia." *Quaternary Science Reviews* 24, 1111–21 (2005).
———. "How Did Humans First alter Global Climate?" *Scientific American* (March 2005): 46–53.
———. "Humans Took Control of Greenhouse Gases Thousands of Years Ago." *Climatic Change* 61 (2003): 261–93.
———. "Orbital Insolation, Ice Volume, and Greenhouse Gases." *Quaternary Science Reviews* 22 (2003): 1597–1629.
Ruddiman, W. F., and M. E. Raymo. "A Methane-based Time Scale for Vostok Ice." *Quaternary Science Reviews* 22 (2003): 141–55.
Ruddiman, W. F., and J. S. Thomson. "The Case for Human Causes of Increased Atmospheric CH_4 over the Last 5000 Years." *Quaternary Science Reviews* 20 (2001): 1769–77.

AFTERWORD

Joos, F., S. Gerber, I. C. Prentice, B. L. Otto-Bleisner, and P. Valdes. "Transient Simulations of Holocene Atmospheric Carbon Dioxide and Terrestrial Carbon since the Last Glacial Maximum." *Global Biogeochemical Cycles* 18 (2004), GB2002, doi: 10.1029/2003GB002156.
Kutzbach, J. E., W. F. Ruddiman, S. J. Vavrus, and G. Phillipon. "Climate Model Simulation of Anthropogenic Influence on Greenhouse-Induced Climate Change (Early Agricultural to Modern): The Role of Ocean Feedbacks." *Climatic Change* (2009), doi: 10.1007/s10584-009-9684-1.
Meure, M., et al. "Law Dome CO_2, CH_4, and NO_2 Ice Core Records Extended to 2000 Years BP." *Geophysical Research Letters* 33 (2006), doi:10.1029/2006GL026152.
Ruddiman, W. F., and A. G. Carmichael. "Pre-Industrial Depopulation Episodes and

Climate Change." In *Global Environmental Change and Human Health*, edited by M. O. Andreae, U. Confalonieri, and A. J. McMichael, 158–94. Vatican City: Pontifical Academy of Sciences, 2006.

Ruddiman, W. F., and E. Ellis. "Effect of Per-Capita Land Use Changes on Holocene Forest Clearance and CO_2 Emissions." *Quaternary Science Reviews* (forthcoming). (An online version is currently available at doi:10.1016/j.quascirev.2009.05.022.)

Ruddiman, W. F., Z. Guo, X. Zhou, H. Wu, and Y. Yu. "Rice Farming and Anomalous Methane Trends." *Quaternary Science Reviews* 27 (2008), doi:10.1016/j.quascirev .2008.03.007.

Vavrus, S., W. F. Ruddiman, and J. E. Kutzbach. "Climate Model Tests of the Anthropogenic Influence on Greenhouse-Induced Climate Change: The Role of Early Human Agriculture, Industrialization, and Vegetation Feedbacks." *Quaternary Science Reviews* 27 (2008), doi:10.1016/j.quascirev.2008.04.011.

FIGURE SOURCES AND CREDITS

Figure 1.1. Adapted from W. F. Ruddiman, "The Anthropogenic Greenhouse Era Began Thousands of Years Ago," *Climatic Change* 61 (2003): 261–93.

Figure 2.1. Adapted from P. B. deMenocal, "Plio-Pleistocene African Climate," *Science* 270 (1995): 53–59.

Figure 2.2. Adapted from A. C. Mix et al., "Benthic Foraminifer Stable Isotope Record from Site 849 [0-5 Ma]: Local and Global Climate Changes," *Ocean Drilling Program Scientific Results* 138 (1995): 371–412.

Figure 3.1. Adapted from W. F. Ruddiman, *Earth's Climate* (New York: W. H. Freeman, 2001).

Figure 3.2. Adapted from Ruddiman, *Earth's Climate.*

Figure 3.3. Adapted from Ruddiman, *Earth's Climate.*

Figure 4.1. Adapted from A. S. Dyke and V. C. Prest, "Late Wisconsinan and Holocene History of the Laurentide Ice Sheet," *Geographie Physique et Quaternaire* 41 (1987): 237–63.

Figure 4.2. Adapted from M. E. Raymo, "The initiation of Northern Hemisphere Glaciation," *Annual Reviews of Earth and Planetary Sciences* 22 (1994): 353–83.

Figure 5.1. Adapted from Ruddiman, *Earth's Climate.*

Figure 5.2. Adapted from COHMAP Project Members, "Climatic Changes of the Last 18,000 Years: Observations and Model Simulations," *Science* 241 (1988): 1043–52.

Figure 5.3. Adapted from W. F Ruddiman and M. E. Raymo, "A Methane-based Time Scale for Vostok Ice," *Quaternary Science Reviews* 22 (2003): 141–55.

Figure 7.1. Adapted from J. Diamond, *Guns, Germs, and Steel* (New York: W. W. Norton, 1997).

Figure 7.2. Adapted from D. Zohary and M. Hopf, *Domestication of Plants in the Old World* (Oxford: Oxford University Press, 1993); and from Diamond, *Guns, Germs, and Steel.*

Figure 8.1. Adapted from T. Blunier et al., "Variations in Atmospheric Methane Concentration during the Holocene Epoch," *Nature* 374 (1995): 46–49; and from Ruddiman, "The Anthropogenic Greenhouse Era."

Figure 9.1. Adapted from Ruddiman and Raymo, "A Methane-based Time Scale for Vostok Ice."

Figure 9.2. Adapted from A. Indermuhle et al., "Holocene Carbon-Cycle Dynamics Based on CO_2 Trapped in Ice at Taylor Dome, Antarctica," *Nature* 398 (1999): 121–26; and from Ruddiman, "The Anthropogenic Greenhouse Era."

Figure 9.3. Adapted from J. W. Lewthwaite and A. Sherratt, "Chronological Atlas," in A. Sherratt, ed., *Cambridge Encyclopedia of Archeology* (Cambridge: Cambridge Press, 1980).

Figure 10.1. Adapted from Ruddiman, "The Anthropogenic Greenhouse Era."

Figure 10.2. Adapted from J. Imbrie and J. Z. Imbrie, "Modeling the Climatic Response to Orbital Variation," *Science* 207 (1980): 943–53.

Figure 10.3. Adapted from Ruddiman, "The Anthropogenic Greenhouse Era."

Figure 10.4. Adapted from W. F. Ruddiman, S. J. Vavrus, and J. E. Kutzbach, "A Test of the Overdue Glaciation Hypothesis," *Quaternary Science Reviews* 24 (2005): 1–10.

Figure 11.1. Adapted from Ruddiman, "The Anthropogenic Greenhouse Era."

Figure 11.2. Adapted from W. F. Ruddiman, "Cold Climate during the Closest Stage 11 Analog to Recent Millennia," *Quaternary Science Reviews* 24 (2005): 1111–21.

Figure 11.3. Adapted from Ruddiman, "Cold Climate."

Figure 11.4. Adapted from Ruddiman, "Cold Climate."

Figure 12.1. Adapted from Indermuhle et al., "Holocene Carbon-Cycle Dynamics."

Figure 12.2. Adapted from H. H. Lamb, *Climate-Past, Present and Future*, vol. 2 (London: Methuen, 1977).

Figure 12.3. Adapted from M. E. Mann, R. S. Bradley, and M. K. Hughes, "Northern Hemisphere Temperatures during the Past Millennium," *Geophysical Research Letters* 26 (1999): 759–62.

Figure 13.1. Adapted from Ruddiman, "The Anthropogenic Greenhouse Era."

Figure 14.1. Adapted from J. T. Andrews et al., "The Laurentide Ice Sheet: Problems of the Mode and Speed of Inception," *Proceedings WMO/IMAP Symposium Publication* 421 (1975): 87–94.

Figure 14.2. Adapted from Ruddiman, "The Anthropogenic Greenhouse Era."

Figure 15.1. Adapted from H. H. Friedli et al., "Ice Core Records of the $^{13}C/^{12}C$ Ratio of Atmospheric CO_2 in the Past Two Centuries," *Nature* 324 (1986): 237–38; and from M.A.K. Khalil and R.A. Rasmusen, "Atmospheric Methane: Trends over the Last 10,000 Years," *Atmospheric Environment* 21 (1987): 2445–52.

Figure 15.2. Adapted from P. A. Mayewski et al., "An Ice-Core Record of Atmospheric Responses to Anthropogenic Sulfate and Nitrate," *Atmospheric Environment* 27 (1990): 2915–19; and from R. J. Charlson et al., "Climate Forcing by Anthropogenic Aerosols," *Science* 255 (1992): 423–30.

Figure 16.1. Adapted from H. S. Kheshgi et al., "Accounting for the Missing Carbon Sink with the CO_2-Fertilization Effect," *Climatic Change* 33 (1996): 31–62.

Figure 16.2. Adapted from Ruddiman, *Earth's Climate.*

Figure 16.3. Adapted from K. Vinnikov et al., "Global Warming and Northern Hemisphere Sea-Ice Extent," *Science* 286 (1999): 1934–39.

Figure 17.1. Adapted from W. F. Ruddiman, "How Did Humans First alter Global Climate?" *Scientific American* (March 2005): 46–53.

Figure A.1. (A) Adapted from L. Loulergue et al., "Orbital and Millennial-Scale Features of Atmospheric CH_4 over the Past 800,000 Years," *Nature* 453 (2008), doi:10.1038/nature 06950. (B) Adapted from D. Luthi et al., "High-Resolution Carbon Dioxide Concentration Record 650,000–800,000 Years before Present," *Nature* 453 (2008), doi:10.1038/nature 06949.

Figure A.3. CO_2 trend adapted from Luthi et al., "High-Resolution Carbon Dioxide Concentration." Populations trends from C. McEvedy and R. Jones, *Atlas of World Population History* (New York: Penguin, 1978); and from W. M. Denevan, *The Native Population of the Americas in 1492* (Madison: University of Wisconsin Press, 1992). Land-use estimates from W. F. Ruddiman and E. Ellis, "Effect of Per-Capita Land Use Changes on Holocene Forest Clearance and CO_2 Emissions," *Quaternary Science Reviews* (forthcoming). (However an online version is currently available at doi:10.1016/j.quascirev.2009.05.022.)

Figure A.4. Adapted from W. F. Ruddiman, Z. Guo, X. Zhou, H. Wu, and Y. Yu, "Rice

Farming and Anomalous Methane Trends," *Quaternary Science Reviews* 27 (2008), doi:10.1016/j.quascirev.2008.03.007.

Figure A.5. (A) Adapted from L. E. Lisiecki and M. E. Raymo, "A Plio-Pleistocene Stack of 57 Globally Distributed Benthic d18O Records," *Paleoceanography* 20 (2005), doi:10.1029/2004PA001071. (B) Adapted from J. Jouzel et al., "Orbital and Millennial Antarctic Climatic Variability over the Past 800,000 Years," *Science* 317 (2007): 793–97.

Figure A.7. Adapted from J. E. Kutzbach, W. F. Ruddiman, S. J. Vavrus, and G. Phillipon, "Climate Model Simulation of Anthropogenic Influence on Greenhouse-Induced Climate Change (Early Agricultural to Modern): The Role of Ocean Feedbacks," *Climatic Change* (2009), doi: 10.1007/s10584-009-9684-1.

Figure A.8. (Top) Adapted from M. Meure et al., "Law Dome CO_2, CH_4, and NO_2 Ice Core Records Extended to 2000 Years BP," *Geophysical Research Letters* 33 (2006), doi:10.1029/2006GL026152. (Bottom) Adapted from McEvedy and Jones, *Atlas of World Population History;* and from Denevan, *The Native Population of the Americas in 1492.*

INDEX